T0227881

Routledge Revivals

U.S. Energy Policies

U.S. Energy Policies, first published in 1968, aims to assemble and describe within an overall framework the energy policy questions that RRF believed would profit from study and analysis. This study covers the past performance and trends in the energy industries, the nature of existing industries and of the government policies bearing on them, and the effects of those policies. This title also takes note of the prospective influence of economic and technological developments and evaluates the probable effects of selected alternatives to existing policies. This book will be of interest to students of environmental studies.

U.S. Energy Policies

An Agenda for Research

A Resources for the Future Staff Report

RFF PRESS
RESOURCES FOR THE FUTURE

First published in 1968
by Resources for the Future, Inc.

This edition first published in 2015 by Routledge
2 Park Square, Milton Park, Abingdon, Oxon, OX14 4RN
and by Routledge
711 Third Avenue, New York, NY 10017

Routledge is an imprint of the Taylor & Francis Group, an informa business

© 1968 Resources for the Future

All rights reserved. No part of this book may be reprinted or reproduced or utilised in any form or by any electronic, mechanical, or other means, now known or hereafter invented, including photocopying and recording, or in any information storage or retrieval system, without permission in writing from the publishers.

Publisher's Note
The publisher has gone to great lengths to ensure the quality of this reprint but points out that some imperfections in the original copies may be apparent.

Disclaimer
The publisher has made every effort to trace copyright holders and welcomes correspondence from those they have been unable to contact.

A Library of Congress record exists under LC control number: 68028767

ISBN 13: 978-1-138-85715-5 (hbk)
ISBN 13: 978-1-315-71885-9 (ebk)
ISBN 13: 978-1-138-85717-9 (pbk)

U.S.ENERGY POLICIES

an agenda for research

A RESOURCES FOR THE FUTURE
STAFF REPORT

RESOURCES FOR THE FUTURE, INC.
Distributed by The Johns Hopkins Press
Baltimore, Maryland 21218

RESOURCES FOR THE FUTURE, INC.
1755 Massachusetts Avenue, N.W., Washington, D.C. 20036

Board of Directors   William S. Paley, Chairman
                       Robert O. Anderson
                       Harrison Brown
                       Erwin D. Canham
                       Edward J. Cleary
                       Joseph L. Fisher
                       Luther H. Foster
                       Charles J. Hitch
                       Edward S. Mason
                       Frank Pace, Jr.
                       Stanley H. Ruttenberg
                       Lauren K. Soth
                       P. F. Watzek

Honorary Directors   Horace M. Albright
                       Reuben G. Gustavson
                       Hugh L. Keenleyside
                       Leslie A. Miller
                       Laurance S. Rockefeller
                       John W. Vanderwilt

President   Joseph L. Fisher
Vice President   Michael F. Brewer
Secretary-Treasurer John E. Herbert

Resources for the Future is a non-profit corporation for research and education in the development, conservation, and use of natural resources. It was established in 1952 with the co-operation of the Ford Foundation and its activities since then have been financed by grants from the Foundation. Part of the work of Resources for the Future is carried out by its resident staff, part supported by grants to universities and other non-profit organizations. Unless otherwise stated, interpretations and conclusions in RFF publications are those of the authors; the organization takes responsibility for the selection of significant subjects for study, the competence of the researchers, and their freedom of inquiry.

RFF staff editors, Henry Jarrett, Vera W. Dodds, Nora E. Roots, Sheila M. Ekers.

© *1968 by Resources for the Future, Inc., Washington, D.C.*
Second printing, 1969
Price: $4.00

Preface

THE REPORT WHICH FOLLOWS was prepared by the staff of Resources for the Future for the Office of Science and Technology, in the Executive Office of the President. It is exclusively the work of RFF and has no official status as a government document. Its purpose is to assemble and describe within an overall framework the energy policy questions which RFF believes would profit from study and analysis. The report is therefore an agenda for research and is not in itself a study that offers energy policy recommendations.

This project began in March 1967 when, at the request of the Office of Science and Technology of the Executive Office of the President, Resources for the Future undertook the design of a comprehensive approach to a study of the energy situation of the United States from the standpoint of public policy. We were glad to accept an assignment of such importance and one so clearly within the field of RFF's interest.

The Office of Science and Technology asked us to design an approach to the study of energy policies as a first step in carrying out the President's assignment of responsibility to that Office to co-ordinate energy policies on a government-wide basis.

In his message to Congress on "protecting our natural heritage" on January 30, 1967, President Johnson said:

> The number and complexity of federal decisions on energy issues have been increasing, as demand grows and competitive situations change. Often decisions in one Agency and under one set of laws—whether they be regulatory standards, tax rules or other provisions—have implications for other Agencies and other laws, and for the total energy industry. We must better understand our future energy needs and resources. We must make certain our policies are directed toward achieving these needs and developing those resources.
>
> I am directing the President's Science Adviser and his Office of Science and Technology to sponsor a thorough study of energy resources. . . .

Resources for the Future was asked to design a broad set of energy studies for consideration of the Office of Science and Technology. As a group, the studies would aim at (1) enlarging understanding, within and outside government, of the characteristics, present position, and prospects of the energy industries; and (2) providing background for policy decisions. The studies would cover the past performance and current trends in the energy industries, the nature of existing industries and of the government policies bearing on them, and the effects—inadvertent as well as intended—of those policies. They would also take note of the present and prospective influence of economic and technological developments and would evaluate the probable effects of selected alternatives to existing policies.

The present volume seeks to back up its recommendations for research with a concise description and analysis of the energy industries in the United States, primarily from the standpoint of policy issues. The book opens with a general view of the energy industries, continues with separate chapters on oil, natural gas, coal, electricity, nuclear energy, and shale oil, and closes with a discussion of the approach to a co-ordinated view of energy problems.

The RFF design study was submitted in final form to the Office of Science and Technology in February 1968. The design study includes much information and raises a multitude of important issues on energy situations and problems in the United States that should be of interest and value to many persons in industry, universities, in national and state government, and in foreign governments and international organizations. Accordingly, OST agreed with RFF on the desirability of making this advisory report publicly available.

Principal members of the RFF group who produced the design study were Sam H. Schurr, director of RFF's program of research in energy and minerals; Hans H. Landsberg, director of RFF's resources appraisal program; and Paul T. Homan, who is a consultant to RFF in its energy research. Temporary consultant Joseph W. Mullen assisted in the project, and other staff members, particularly Joel Darmstadter, contributed to specific aspects of the work. Several of RFF's published studies were drawn upon[1] along with some still unpublished materials. Among the latter, the as yet unpublished manuscript of the Louis Lister–Paul Homan study of government energy policies in the United States was particularly helpful.

We also are grateful to the persons in industry and government and in

---

[1] Notably *Economic Aspects of Oil Conservation Regulation*, by Wallace F. Lovejoy and Paul T. Homan; *Resources in America's Future*, by Hans H. Landsberg, Leonard L. Fischman, and Joseph L. Fisher; and *Energy in the American Economy, 1850-1975*, by Sam H. Schurr and Bruce C. Netschert, with Vera F. Eliasberg, Joseph Lerner, and Hans H. Landsberg (all three published by The Johns Hopkins Press); and *Energy in the United States*, by Hans H. Landsberg and Sam H. Schurr (Random House).

universities and other academic institutions and in trade associations, who provided ideas for the design study and reviewed and commented upon all or parts of the manuscript while the study was in progress. The names —about seventy-five—are too numerous to list here, but our debt is none the less great.

Washington, D.C. Joseph L. Fisher, President
May, 1968 Resources for the Future, Inc.

Contents

Tables

U.S. Energy Policies:
An Agenda for Research

I

A General View
of the Energy Industries

"ENERGY" IS A GENERIC WORD of many applications and nuances, associated with the general idea of "capacity to do work." For purposes pertinent to an economic study, it may be confined mainly to functions related to activating mechanical equipment and providing heat in other useful ways.

Conventional terminology distinguishes between primary and secondary sources. The primary sources of energy may be listed as crude oil, natural gas, natural gas liquids, coal, and falling water. Most recently, nuclear fuels have been added to the list as economically usable sources, and, in central Canada, the first oil from tar sands began to flow in September 1967. On the horizon of economic utilization are great deposits of oil-bearing shale in the western United States. Primary sources, such as geothermal and solar, are not judged to be of sufficient weight in the foreseeable U.S. energy mix to raise policy issues calling for research in the next few years. They are, therefore, not dealt with in this study.

Electricity is classified as a "secondary" source of energy, because the primary sources are used to produce it. Once produced, it performs various services that are unique to it; but it also performs some services of the same sort as those provided by the primary sources. Two of the primary sources, water and nuclear fuels (as presently applied), are most closely related to electricity in that their current economic usefulness in the context of energy is confined almost solely to the generation of electricity. Coal has a close relationship in that a major fraction of all coal produced in the United States is used for electrical generation. Oil and natural gas are also utilized for producing electricity, but this service constitutes only a minor portion of their uses.

1

The various energy sources are substitutable for one another over a wide range. Railway locomotives may be fired by oil or coal, or run by electricity. Buildings may be heated with coal, oil, natural gas, or electricity. The boiler fuel to create steam pressure for process heat or for activating industrial equipment may be coal, oil, or natural gas. Industrial furnaces for processes that do not require steam may derive heat from fossil fuels or from electricity.

The extent to which particular forms of energy are applied to particular uses is in part the result of changing supply conditions and prices of the various sources; in part, it is dependent on changing technologies which establish preferential efficiencies in various uses. In some cases a single source of energy will entirely displace another; oil, for example, has replaced coal as locomotive fuel. Electricity has replaced oil and gas for lighting purposes. More commonly, however, two or three of the sources are in use at the same time for the same purposes, as for space heating and industrial boiler fuel. The margins of use are established by economic cost-price factors, and frequently contain a geographical factor. For most purposes high transport costs allow coal to compete only exceptionally in areas where oil and gas are produced; but oil and gas can penetrate coal-bearing regions.

Though interchangeability is the rule, in some uses single sources of energy take over the whole field, mainly for technological reasons but partly for reasons of cost. Oil provides the sole fuel for internal combustion motors, though other fuels could be used and other means than internal combustion employed for mobile vehicles. For a variety of industrial processes, the electrical motor is the sole applicable agent.

Because the sources of energy are so generally interchangeable, for many planning and policy purposes energy must be considered in the aggregate. In all projections of economic growth the future requirements for energy expand significantly; much faster than population, but not quite as fast as gross national product. Public policy is necessarily concerned, among other things, with assuring the availability of expanded supplies adequate to these needs. This concern often need not take the form of separate concern for supplies from individual sources—gas, oil, coal, atomic energy, etc.—but with their sum, whatever the sources. But the sum is, of course, made up of the parts; and thought must be given to the contribution of the parts severally.

Public Involvement with the Energy Industries

In approaching the subject of energy from this collective point of view, we must first pay some attention to the forms of the separate industries that supply energy. Each industry has developed its own characteristic form. The business firms in each industry are in competition with one another for their place in the market. At the same time, the members of each industry are in competition with other industries, all of which are attempting to expand their positions in the total energy market. Oil producers and processors, for example, are in competition for their share in the market for oil products; but they are also in competition with natural gas producers and coal producers for those markets wherein their products are interchangeable as energy sources.

The method of relying on private enterprise, not only to establish for each firm its position in its own industry but also to establish the role of the industry in the total economy, is the primary characteristic of the American economic system. This method is normally carried out under the provisions of the antitrust law policy designed to prevent monopolistic practices, except for a few special industries of the public utility type where industries take the form of total or partial monopolies subject to public regulation. To the extent that the energy industries conform to these characteristic patterns of industrial organization, they are simply normal members of the American economic universe. The fact is, however, that the separate energy industries have given rise to special problems which bring all of them, except coal, under special, and diverse, methods of direct public regulation—and coal also has been regulated in the past.

In the production of *crude oil*, various state governments regulate the amount of production in the state and prorate the total into production quotas for individual producers. In sum, these state regulatory practices fix the total supply of oil produced in the nation and tend to dampen price fluctuations. Through its control of the Continental Shelf, the federal government is involved in leasing lands for development of the oil and gas provinces there located. In addition, on the ground that an undue increase would endanger security, the federal government has since 1959 limited oil imports, thereby reinforcing the controls exercised by the states.

In the case of *natural gas*, in addition to a degree of special state regulation the industry is regulated at a series of levels by state and federal agencies. At the local consumer level, prices and services are regulated by state public utility commissions. At the level of interstate transmission, the wholesale prices and services of the pipeline companies are regulated by the federal government. At the level of primary production, the field prices paid by interstate pipelines to producers are regulated by the federal government.

In the case of *coal*, there is no direct public regulation; prices and production are left to the forces of the competitive market. But the costs of delivered coal are substantially affected by railroad rates which are regulated by the federal government and have an important effect on the terms of competitive entrance of producers into local markets. A rather special type of agreement with the union also strongly affects the competitive situation. The federal government provides some assistance through support of a program of market and technical research.

In the case of *electricity*, local rates and services of private companies are regulated by public utility commissions in most states. Under federal law, the wholesale price of electricity in interstate commerce is subject to federal regulation. The federal government is a direct producer and wholesaler of electricity from numerous hydroelectric installations in multipurpose river basin development programs; and through the Tennessee Valley Authority it is the sole supplier to a wide geographical region. It licenses private companies to use interstate rivers for hydroelectric projects. Some states operate hydroelectric projects. The federal government also supports rural electric co-operatives through low-interest loans and assists them in other ways. Many municipalities own and operate their own generating facilities, and in the state of Nebraska only publicly and co-operatively owned utility systems are allowed.

The preceding paragraphs, without being exhaustive, suggest some of the types of government involvement in the operation of the various energy industries. This involvement becomes more extensive when concern is with new sources of energy now on the economic and technological horizon, and tends to bring into focus especially issues associated with size and allocation of federal funds for research and development.

Nuclear energy has just crossed the border of economic usefulness for electrical generation with the aid of a vast federal expenditure on research and development. In 1966 and 1967, more than 50 per cent of the new electricity generating capacity ordered, or announced for future construction, consisted of nuclear power plants. The present technology is, however, only a beginning. To bring out the tremendous energy content of nuclear sources will require extended additional research and development. As nuclear power originated in the hands of government, many elements of control over the introduction of nuclear processes into the energy scene have rested with government, and some still do, even though actual productive organization has been supplied by private enterprise.

In another direction, *shale oil*, an extended involvement of government in future energy supplies is developing. It stems primarily from the fact that the bulk of the rich deposits of oil shale, containing an oil content vastly greater than the probable content of the petroleum reservoirs, are located in publicly owned lands in the Rocky Mountain region. Though research and development to bring them within the economic horizon has heretofore been meager and desultory, there is little doubt that they will ultimately provide a large addition to available energy supplies. Government policy will play a large part in determining the course of research

and development as well as the pace and the nature of the industrial organization and terms under which the shale resources are exploited.

We see, then, that government policies have actively affected the
operations of the existing energy industries. There is, however, no common basis among the policies applicable to the several industries other
than general encouragement for development of additional resources, and
even on this, one might not receive unanimous agreement. No "energy
policy" as such emerges from the picture. The policies relating to each industry have arisen out of problems encountered in that industry, and
have been shaped by a variety of technical, economic, and ideological
factors in changing political situations.

Postwar Growth and Composition of Energy

Between 1947 and 1965, consumption of energy in the United States
rose by an annual average of 2.8 per cent compounded, just a shade below the long-term growth rate of about 3 per cent.[1] Although it rose in all
but five of these eighteen years, the rate of increase was markedly below
the 2.8 per cent average in the first few years of the period (which followed an expansion in energy use unprecedented since the beginning of
the century) and especially fast in the first half of the 1960's.

During the same two decades population rose by 1.7 per cent per year,
and gross national product, in real terms, by 3.9 per cent annually. Thus,
energy consumption followed the historical pattern of rising substantially
faster than population. At the same time, the trend of falling energy consumption per dollar of GNP, which had set in at the end of World War I,
persisted at essentially the same rate between 1947 and 1965—i.e., an
average decline in the energy-GNP ratio of about one per cent yearly
(Table 1).

Steady growth in energy consumption was accompanied by significant
changes in the energy mix, as expressed calorically and shown in Table 2.
Coal ceased to be the dominant source of energy, being surpassed at the
beginning of the 1950's by oil and, less than a decade later, by natural
gas.

By the mid-1960's these changes seemed to be leveling off, and an ap-
proximate pattern had emerged in which oil held 40 per cent, gas 30 per

[1] The common denominator here used to sum up and compare the several energy
sources is the British thermal unit (Btu), a calorific unit representing the amount
of heat necessary to raise the temperature of one pound of water one degree Fahren-
heit. This usage is not directly applicable to water power used for electrical generation, which has been measured by applying the heat content of an alternative thermal
source for producing the same amount of electricity. The Btu common denominator
makes no allowance for the relative efficiency in the use of the source materials
for various purposes, either as against one another at a given time, or through
time under changing technologies.

cent, and coal a little over 20 per cent of the domestic energy market—a far cry from 1947 when the corresponding figures were 33, 14, and 44.

Over the same period, the share of anthracite had declined from 4 to less than 1 per cent, that of hydropower from about 4½ to a little under 4 per cent, and natural gas liquids had risen from 1½ to 3½ per cent,

TABLE 1. Data on U.S. Energy Consumption, Population, and Gross National Product, Selected Years, 1900-1965

Year	Energy consumption (*trillion Btu*)	Population (*thousands*)	GNP (*billion 1958 dollars*)	Energy consumption per capita (*million Btu*)	Energy consumption per dollar of GNP (*thous. Btu*)	Indexes (1947=100) of columns: (1)	(4)	(5)
	(1)	(2)	(3)	(4)	(5)	(6)	(7)	(8)
1900	7,572	76,094	79.8	99.5	94.9	23.0	43.8	89.4
1905	11,369	83,820	100.0	135.6	113.7	34.6	59.7	107.2
1910	14,800	92,407	120.1	160.2	123.2	45.0	70.5	116.1
1915	16,076	100,549	124.5	159.9	129.1	48.9	70.4	121.7
1920	19,782	106,466	140.0	185.8	141.3	60.2	81.8	133.2
1925	20,899	115,832	179.4	180.4	116.5	63.6	79.4	109.8
1930	22,288	123,077	183.5	181.1	121.5	67.8	79.7	114.5
1935	19,107	127,250	169.5	150.2	112.7	58.1	66.1	106.2
1940	23,908	132,594	227.2	180.3	105.2	72.7	79.4	99.2
1945	31,541	140,468	355.2	224.5	88.8	96.0	98.8	83.7
1947	32,870	144,698	309.9	227.2	106.1	100.0	100.0	100.0
1950	34,153	152,271	355.3	224.3	96.1	103.9	98.7	90.6
1953	37,697	160,184	412.8	235.3	91.3	114.7	103.6	86.0
1955	39,956	165,931	438.0	240.8	91.2	121.6	106.0	86.0
1957	41,922	171,984	452.5	243.8	92.6	127.5	107.3	87.3
1960	44,816	180,684	487.8	248.0	91.9	136.3	109.2	86.6
1961	45,573	183,756	497.2	248.0	91.7	138.6	109.2	86.4
1962	47,620	186,656	529.8	255.1	89.9	144.9	112.3	84.7
1963	49,649	189,417	551.0	262.1	90.1	151.0	115.4	84.9
1964	51,515	192,120	580.0	268.1	88.8	156.7	118.0	83.7
1965	53,791	194,572	614.4	276.5	87.6	163.6	121.7	82.6

Sources: Energy consumption, 1900-1955, as shown in U.S. Bureau of the Census, *Historical Statistics of the United States, Colonial Times to 1957* (Washington: Government Printing Office, 1960), Series No. M78 plus M85, pp. 354-55; 1957-65, from U.S. Bureau of Mines, *Minerals Yearbook*, Vol. II, *Fuels* (Washington: Government Printing Office, various issues). Population data from U.S. Bureau of the Census, *Statistical Abstract of the United States, 1966* (Washington: Government Printing Office, 1966), p. 5. GNP data from U.S. Bureau of the Census, *Long Term Economic Growth, 1860-1965* (Washington: Government Printing Office, 1966), pp. 166-67. (The 1910-65 GNP figures are U.S. Commerce Department estimates; the linked 1900 and 1905 estimates are based on a series of John W. Kendrick shown in *Long Term Economic Growth, 1860-1965*.)

TABLE 2.   U.S. Energy Consumption, by Primary Fuel,
Selected Years, 1900-1965

Year	Total	Bituminous coal [a]	Anthra-cite	Crude oil [b]	Natural gas	Natural gas liquids	Hydro-electric power [c]
			A. TRILLION BTU				
1900	7,572	5,431	1,410	229	252	–	250
1910	14,800	10,654	2,060	1,007	540	–	539
1920	19,782	13,325	2,179	2,634	827	42	775
1925	20,899	13,079	1,627	4,156	1,212	124	701
1930	22,288	11,921	1,718	5,652	1,969	246	785
1940	23,908	11,290	1,245	7,487	2,726	294	917
1945	31,541	14,661	1,311	9,619	3,973	491	1,486
1947	32,870	14,302	1,224	10,803	4,518	564	1,459
1950	34,153	11,900	1,013	12,706	6,150	783	1,601
1955	39,956	11,104	599	16,328	9,232	1,196	1,497
1960	44,816	9,967	447	18,608	12,736	1,427	1,631
1961	45,573	9,809	404	18,989	13,228	1,498	1,645
1962	47,620	10,160	363	19,662	14,027	1,605	1,803
1963	49,649	10,722	361	20,282	14,843	1,668	1,773
1964	51,515	11,295	365	20,617	15,562	1,769	1,907
1965	53,791	12,030	328	21,337	16,136	1,872	2,088
			B. PERCENTAGE DISTRIBUTION				
1900	100.0%	71.7%	18.6%	3.0%	3.3%	–	3.3%
1910	100.0	72.0	13.9	6.8	3.7	–	3.6
1920	100.0	67.4	11.0	13.3	4.2	0.2%	3.9
1925	100.0	62.6	7.8	19.9	5.8	0.6	3.3
1930	100.0	53.5	7.7	25.4	8.8	1.1	3.5
1940	100.0	47.2	5.2	31.4	11.4	1.0	3.8
1945	100.0	46.5	4.2	30.5	12.6	1.5	4.7
1947	100.0	43.5	3.7	32.9	13.8	1.7	4.4
1950	100.0	34.8	3.0	37.2	18.0	2.3	4.7
1955	100.0	27.8	1.5	40.8	23.1	3.0	3.8
1960	100.0	22.0	1.0	41.6	28.4	3.2	3.6
1961	100.0	21.5	0.9	41.6	29.0	3.3	3.7
1962	100.0	21.3	0.8	41.3	29.4	3.4	3.8
1963	100.0	21.6	0.7	40.8	29.9	3.4	3.6
1964	100.0	21.9	0.7	40.0	30.2	3.5	3.7
1965	100.0	22.4	0.6	39.6	30.0	3.5	3.9

[a] Includes lignite.

[b] Includes net imports of petroleum products.

[c] Includes nuclear for 1960-65 (never exceeding 0.1 per cent). For approach used to measure hydro and nuclear, see Table 4, footnote h.

Sources: Same as sources cited for energy consumption in Table 1.

TABLE 3. U.S. Production of Fuel and Non-Fuel Minerals (Quantities and Value), 1947, 1955, 1965

Item	1947				1955				1965			
			Value — As per cent of—				Value — As per cent of—				Value — As per cent of—	
	Quantity	Dollars (mill.)	All minerals	Fuels only	Quantity	Dollars (mill.)	All minerals	Fuels only	Quantity	Dollars (mill.)	All minerals	Fuels only
Coal, total (million short tons)	687	3,034	31.6%	42.2%	490	2,295	14.5%	21.3%	528	2,404	11.2%	17.1%
Bituminous coal and lignite	630	2,620	27.3	36.5	464	2,087	13.2	19.4	512	2,276	10.6	16.2
Pennsylvania anthracite	57	413	4.3	5.7	26	206	1.3	1.0	15	122	0.6	0.9
Peat	0.1	1	...	...	0.3	2	...	...	0.6	6	...	...
Natural gas, wet (billion cu. ft.)	4,582	275	2.9	3.9	9,405	978	6.2	9.1	16,040	2,495	11.6	17.8
Natural gas liquids (million gallons)	132	295	3.1	4.1	11,818	619	3.9	5.8	18,545	911	4.3	6.5
Natural gasoline and cycle products	87	228	2.4	3.2	5,845	424	2.7	4.0	7,288	494	2.3	3.5
L-P gases	45	67	0.7	0.9	5,973	195	1.2	1.8	11,257	417	2.0	3.0
Petroleum, crude (million barrels)	1,857	3,578	37.2	49.8	2,484	6,871	43.6	63.8	2,848	8,158	38.1	58.1
Total, fuels [a]	b	7,188	74.8	100.0	b	10,774	68.3	100.0	b	14,015	65.5	100.0
Total, non-fuel minerals	b	2,422	25.2		b	5,003	31.7		b	7,388	34.5	
Metals	b	1,084	11.3		b	2,044	13.0		b	2,472	11.5	
Non-metals	b	1,338	13.9		b	2,959	18.7		b	4,916	23.0	
All minerals	b	9,610	100.0		b	15,777	100.0		b	21,433	100.0	

... Negligible.

[a] Excluding hydro and nuclear power. Data for 1955 and 1965 include small amounts of other mineral energy resources, not shown separately above.

[b] Common weight or volume measure not meaningful.

Source: U.S. Bureau of Mines, Minerals Yearbook, various issues.

much of the material in this category being used not as energy but as feedstocks in chemical manufacturing.

These broad compositional changes in energy use are reflected also in data on trends in the value and volume of fuels output (hydro excluded) over the postwar period, shown in Table 3. However, largely because of differential price movement (to a lesser extent, because of the influence of foreign trade, which still plays a relatively modest part in the U.S. energy economy), the value shares of particular fuels in total fuel output exhibit a somewhat different pattern from the percentages shown in Table 2. Mineral fuels are also seen to make a dominant, though declining, contribution to the value of *total* minerals production.

Whether the reversals of the declining coal and the rising oil trend, and the flattening of the rising gas trend—all in terms of market shares —are steps leading to another radical change in the energy mix is a subject of legitimate controversy, made the more intriguing by the advent of commercial nuclear energy and the approach to competitiveness of other hitherto unused energy sources.

It is, therefore, important to understand the past. What, one asks, are some of the events that underlie these recent changes? In the case of coal, these consist of the loss of the railroad and space heating—and to a lesser extent, industrial—market in the early stages of the period and the subsequent boost it received from its traditional tie to the rapidly expanding thermal electric power industry. Peak production in coal's second-best customer, the steel industry, gave an added stimulus in the early 1960's. Thus the slide of several decades leveled off at about 21 per cent in 1962, and between 1962 and 1965 the share of coal in energy consumption actually increased somewhat. Though minor, the increase marks the first time since World War II that the decline in coal's percentage has been halted for more than one year at a time.

Both oil and gas gained from coal's loss. Oil picked up coal's losses in railroad transportation, and both oil and gas moved heavily into the space heating market. In addition, gas made rapid gains as a boiler fuel, especially in electric power generation, and thus limited coal's gains in this application. Simultaneously, gas made heavy inroads on oil, especially in the residential heating market. With its space heating market curtailed and its participation in the growth of electric power generation minor, oil's relative stability derives above all from its unchallenged role as a source of energy in transportation. By the same token, its future appears closely tied to that market.

The growth in the volume of energy consumed in the American economy and changes in the mix have been associated with changes in the importance of various energy-consuming sectors and, within these sectors, by changes in the mix of fuels consumed in them.

TABLE 4. U.S. Consumption of Energy Resources by Major Sources and Consuming Sectors, 1947, 1955, 1965

Consuming sectors	Anthracite	Bituminous coal	Natural gas	Petroleum [a]	Hydro- and nuclear power [b]	Total primary energy inputs [c]	Utility electricity purchased [d]	Total sector energy inputs [e]
				A. TRILLION BTU				
Household and commercial								
1947	813	2,586	1,125	2,251	f	6,774	393	7,167
1955	331	1,444	2,850	4,001	f	8,625	837	9,462
1965	79	546	5,534	5,634	f	11,793	1,937	13,730
Industrial								
1947	285	7,014	2,875	2,490	f	12,663	452	13,115
1955	53	5,796	4,675	3,329	f	13,853	1,000	14,853
1965	49	5,640	7,685	4,141	f	17,515	1,634	19,149
Transportation [g]								
1947	24	3,006	—	5,760	f	8,791	24	8,815
1955	12	462	254	9,109	f	9,837	20	9,857
1965	—	19	518	12,184	f	12,721	21	12,742
Electric utilities [h]								
1947	90	1,994	386	468	1,459	4,397		
1955	82	3,402	1,194	512	1,497	6,686		
1965	55	5,825	2,399	743	2,088	11,110		
Miscellaneous and unaccounted for								
1947	13	—	133	398	f	544		
1955	123	—	260	572	f	955		
1965	145	—	—	507	f	652		

Total primary energy inputs						
1947	1,224	14,600	4,518	11,367	1,459	33,168 i
1955	599	11,104	9,232	17,524	1,497	39,956
1965	328	12,030	16,136	23,209	2,088	53,791

B. PERCENTAGE DISTRIBUTION: CONSUMING SECTORS BY PRIMARY ENERGY SOURCES

Household and commercial						
1947	12.0%	38.2%	16.6%	33.2%	f	100.0%
1955	3.8	16.8	33.0	46.4	f	100.0
1965	0.7	4.6	46.9	47.8	f	100.0
Industrial						
1947	2.2	55.4	22.7	19.7	f	100.0
1955	0.4	41.8	33.7	24.1	f	100.0
1965	0.3	32.2	43.9	23.6	f	100.0
Transportation g						
1947	0.3	34.2	—	65.5	f	100.0
1955	0.1	4.7	2.6	92.6	f	100.0
1965	—	0.1	4.1	95.8	f	100.0
Electric utilities h						
1947	2.1	45.3	8.8	10.6	33.2%	100.0
1955	1.2	50.9	17.8	7.7	22.4	100.0
1965	0.5	52.4	21.6	6.7	18.8	100.0
Miscellaneous and unaccounted for						
1947	2.4	—	24.4	73.2	f	100.0
1955	12.9	—	27.2	59.9	f	100.0
1965	22.2	—	—	77.8	f	100.0
Total primary energy inputs						
1947	3.7	44.0	13.6	34.3	4.4	100.0
1955	1.5	27.8	23.1	43.9	3.7	100.0
1965	0.6	22.4	30.0	43.1	3.9	100.0

(See footnotes on p. 13)

TABLE 4 continued

C. PERCENTAGE DISTRIBUTION: PRIMARY ENERGY SOURCES BY CONSUMING SECTORS

Consuming sectors	Primary energy sources						Utility electricity purchased [d]	Total sector energy inputs [e]
	Anthracite	Bituminous coal	Natural gas	Petroleum [a]	Hydro- and nuclear power [b]	Total primary energy inputs [c]		
Household and commercial								
1947	66.4%	17.7%	24.9%	19.8%	[f]	20.4%		
1955	55.2	13.0	30.9	22.8	[f]	21.6		
1965	24.1	4.5	34.3	24.3	[f]	21.9		
Industrial								
1947	23.3	48.0	63.6	21.9	[f]	38.2		
1955	8.7	52.2	50.6	19.0	[f]	34.7		
1965	14.9	46.9	47.6	17.8	[f]	32.6		
Transportation [g]								
1947	1.9	20.6	—	50.7	[f]	26.5		
1955	2.0	4.2	2.8	52.0	[f]	24.6		
1965	—	0.2	3.2	52.5	[f]	23.6		
Electric utilities [h]								
1947	7.4	13.7	8.6	4.1	100.0%	13.3		
1955	13.6	30.6	12.9	2.9	100.0	16.7		
1965	16.8	48.4	14.9	3.2	100.0	20.7		
Miscellaneous and unaccounted for								
1947	1.0	—	2.9	3.5	[f]	1.6		
1955	20.5	—	2.8	3.3	[f]	2.4		
1965	44.2	—	—	2.2	[f]	1.2		

	1947	1955	1965	1947	1955	1965	1947	1955	1965	1947	1955	1965

Total primary energy
inputs
1947   100.0   100.0   100.0   100.0   100.0
1955   100.0   100.0   100.0   100.0   100.0
1965   100.0   100.0   100.0   100.0   100.0

a Including natural gas liquids.

b Includes nuclear for 1965 only (negligible).

c Represents energy content of all primary energy sources and their derivatives (domestic and imported) at time it is incorporated in the indicated consuming sectors of the economy. Hydro- and nuclear power, where entered, are converted at their theoretical primary energy equivalent (see footnote h, below).

d Represents utility electricity distributed to the indicated consuming sectors, computed at 3,412 Btu per kwh, the direct calorific value of electricity.

e Energy resource inputs by sector, including direct fuels and electricity purchases.

f Not shown, since only aggregate electric energy purchases can be allocated to end uses.

g Comprises all transportation, including commercial transport services, passenger cars, bunkers, and military.

h The hydro- and nuclear power component of this panel are converted to theoretical inputs at the prevailing average rate of pounds of coal per kwh at central electric stations.

i Owing to revision, differs slightly from corresponding figure in Tables 1 and 2.

Note: A dash signifies "nil," "not applicable," or "negligible."

Source: All data from U.S. Bureau of Mines, *Minerals Yearbook, 1963* and *1965*, Vol. II, *Fuels*, except 1947 and 1955 "utility electricity purchased," which was obtained from W. A. Vogely and W. E. Morrison, "Patterns of Energy Consumption in the United States," paper presented at the World Power Conference, Tokyo, October 1966.

In order to bring out one of the outstanding changes during the two decades, it is convenient to treat electric power generation as a separate economic sector rather than as a processing stage prior to final consumption. The results are shown in Table 4, in which the relative rise in energy use by the electric utility industry is seen to be the steepest of all. Even so, this rise is understated in an important sense, since, being expressed in terms of the energy *input* required to generate the electricity, it is restrained by the steady improvement in the utility industry's thermal efficiency. This is easily seen when electric power output is compared with the input of energy resources into the industry. In that comparison output rises by about 310 per cent from 1947 to 1965, as reflected in the sum of purchases from utilities in Table 4, and input by 150 per cent. Even the input, however, is impressive when compared with a rise of 75 per cent in primary energy consumption by the household and commercial segment, approximately 45 per cent in transportation, and, least of all, not quite 40 per cent in the industrial segment. The market shares shown in panel C of Table 4 are the results of these movements.

It should be noted that the categories used are exceedingly broad and each cover a multitude of different activities. They thus give few clues to the changing circumstances that result in the differential rates of change we have cited. Defense uses, agriculture, and use of energy materials for non-energy purposes, are all included in one or the other of the three specifically designated economic segments. Compounded with ambiguities in the data as originally collected and classified, these categorizations should not be looked at as more than rough suggestions of change: a very rapid growth in electric power generation, followed far down the line by growth in energy input into household and commercial, transportation, and industrial use, in that order, in terms of rate of increase.

Nonetheless, industrial use has remained the largest customer. It accounts for a third of energy input; followed by transportation, with not quite a quarter; and household and commercial, with about 22 per cent.

When input into power generation is allocated to final consumers rather than considered as an end-use in itself (Table 5), industry remains the largest consumer, but because transportation is for all practical purposes out of the electric power market, it falls considerably below the household and commercial segment as a destination of energy use. Because households are taking a rising proportion of electricity sales, the sectoral energy shares—so conceived—show a fairly marked increase for households, with modestly declining percentages for the other sectors. The industrial share, however, declines less than in Table 4, where electricity is treated as a final consuming sector. All the same, these are not major changes when one considers the roughness of the estimates.

As shown in Table 4, changes in the mix of fuels consumed within the

TABLE 5.   U.S. Energy Consumption, by Broad Use Classification,[a]
Electricity Allocated to Final Consuming Sectors,[b] 1947, 1955, 1965

	1947	1955	1965
	Per cent of total primary energy inputs		
Household and commercial	26.4%	29.1%	33.1%
Industrial	45.1	43.7	42.0
Transportation	26.9	24.8	23.8
Miscellaneous and un- accounted for	1.6	2.4	1.2
Total	100.0%	100.0%	100.0%
Total trillion Btu	33,168	39,956	53,791

[a] As the classifications suffer from several defects, breakdowns should be viewed as broad approximations, e.g., commercial use includes an unknown amount of consumption by small industrial users of gas and electricity, though presumably not of coal and oil; transportation is deficient in coverage of some water-borne traffic.

[b] The primary energy inputs (including theoretical primary equivalent of hydro- and nuclear power) of electric utilities have been allocated according to shares of sectors of utility electricity sales.

Source: Table 4.

sectors have been more pronounced than changes in sectoral shares of total energy consumption. In household and commercial use, oil and gas have virtually eliminated direct burning of coal; and natural gas, supplying only half as much energy as petroleum in 1947, has advanced to virtual equality, and probably to more than that if one were to exclude natural gas liquids from the petroleum column. The role of coal has thus become an indirect one, operating through coal-fired power plants that supply electricity to homes and commercial establishments. But even when coal so applied is considered at its caloric input value, total coal use in this segment of the economy still lagged substantially behind both gas and oil in the mid-1960's.

In the industrial sector, too, gas and to a lesser extent oil have grown at the expense of coal, though not to the degree that this has occurred in household and commercial use. In 1965, each of the fuels retained significant shares in industrial consumption of energy, largely because of coal's firm roots in the metal industry and a few other large industrial consumers. The sharp leveling off from coal's absolute decline, between 1955 and 1965, is notable.

In the transportation industry, oil, which shared the market with coal on a 2:1 basis at the end of World War II, has almost pre-empted it. The loss to oil of the rail market had been completed by the mid-1950's, and the rapid expansion in road and air transport markets favored oil, not coal. Thus there was no cushion in the transportation market that could soften coal's decline.

TABLE 6. Data on Costs and Use of Fuels at U.S. Thermal Electric Power Stations, Selected Years, 1948-1965

Year	Cost (cents per million Btu)				Fuel cost per kwh of electricity (in mills)	Thermal electricity generated using— (Index: 1948 = 100)			
	Coal	Oil	Gas	All fuels		Coal	Oil	Gas	All fuels
	(1)	(2)	(3)	(4)	(5)	(6)	(7)	(8)	(9)
1948	27.9	45.3	10.4	26.7	4.2	100	100	100	100
1955	26.1	38.3	18.7	24.3	2.8	197	222	317	217
1960	26.0	34.3	24.4	26.2	2.8	263	275	525	304[a]
1965	24.4	33.3	24.9	25.2	2.6	373	387	736	430[a]

[a] Includes small quantities of nuclear generation.

Sources: Cols. (1) through (5) for all years, and cols. (6) through (9) for 1955-65 from Edison Electric Institute, Historical Statistics of the Electric Utility Industry (New York: Edison Electric Institute, 1962), p. 34; and Statistical Yearbook of the Electric Utility Industry for 1965 (New York: Edison Electric Institute, 1966), pp. 18, 46. Cols. (6) through (9) for 1948 derived from U.S. Bureau of the Census, Historical Statistics of the United States, Colonial Times to 1957, Series S38-S43, p. 507.

In power generation, oil and hydropower have both lost ground and nuclear energy in 1965 was contributing only a tenth of one per cent. It is, however, the only sector in which coal consumption has increased both absolutely and relatively; in 1965, over 50 per cent of utility power was coal derived, compared with 45 per cent in 1947. The one great change in sources of energy for this sector has been the entry of natural gas. The rise of natural gas as a boiler fuel has its root in the rapid expansion of the gas pipeline network, the introduction of gas sales to power plants on an interruptible basis and therefore at low rates, and rapid economic growth in the southwestern part of the United States, near gas-producing areas. They all supported an increase in the share of natural gas from 9 per cent in 1947 to 22 per cent in 1965.

Much of the change in energy sources for power generation is rooted in the cost of fuel to utilities, shaped in turn by changes in technology. Pipelines and pricing policies made natural gas available in increasing parts of the country at rates that compared favorably with those of coal and oil (Table 6). Slowly but steadily gas prices rose, and by the early 1960's they had, on the national average, come to equal coal prices. But in the process, with continuous improvements in thermal efficiency, the average fuel cost per kilowatt-hour generated had declined from 4.2 mills in 1948 to 2.6 mills in 1965. The seeming paradox that the role of natural gas expanded substantially while its price was rising thus merely reflects the initial low price of gas in terms of its energy content. This was partly due also to lower fixed charges, for in the temperate climate of gas-producing areas in the Southwest plants can be built more cheaply than farther north, and there is no need for coal and ash-handling equipment. The need to manage environmental pollution is likely to add further capital charges to coal-burning plants.

Geography has influenced events in other ways. Although little has happened to diminish the importance of the Appalachians as the center of U.S. coal production during the past two decades, the 1960's might be remembered as the decade in which western coal, in New Mexico, Arizona, and Utah, began to come into its own, supporting or about to support large electric power plants in those regions. In the case of oil, Louisiana has made the strongest advance during the postwar period, overtaking first Oklahoma and most recently California to become second only to Texas among petroleum-producing states. Another change has been the emergence of the Mountain states as important oil producers. As for natural gas, Texas retains the dominant position, though Louisiana has been rising fast.

Foreign Trade

The postwar period witnessed the transition of the country's energy economy from net exporter to net importer. The year 1949 marks the turning point. Since that time the deficit in the country's energy trade account has steadily risen, held down since the late 1950's only by restrictions on oil imports. In 1965, net imports of energy resources had risen to 7½ per cent of energy consumption, compared with an export surplus of 6 per cent in 1947 (Table 7). On a gross basis, mineral fuel imports constituted over 10 per cent of total merchandise imports in 1965, with petroleum imports representing about nine-tenths of energy imports and natural gas the balance. The role of imported energy since 1947 is most sharply revealed by its contribution to the growth in energy consumption. Net imports of fuel of all kinds, raw and refined, supplied 34 per cent of the growth in U.S. energy consumption between 1947 and 1955, and 39 per cent between 1955 and 1965.

Gross exports of energy consist almost solely of coal and refined petroleum products—of equal importance when viewed in value terms; calorically, coal exports are substantially more important. Exports of crude oil have, except in abnormal situations, stopped altogether. The value of all gross mineral energy exports was only about 3½ per cent of total merchandise exports in 1965. Coal exports, upon which great hopes had been placed, have remained substantial but have failed to bear out more optimistic predictions, which were based above all on greatly increased shipments to Western Europe. Moreover, steam coal, which used to form the predominant portion of coal exports, is now of secondary importance. Whether coking coal exports can maintain even current levels

TABLE 7.   U.S. Net Energy Imports [a] Related to Energy Consumption, 1947, 1955, 1965

	1947	1955	1965
	(. . . . . trillion Btu . . . .)		
Crude oil	292	1,529	2,517
Petroleum products	−262	372	2,452
Natural gas	−19	−21	437
Bituminous coal and lignite	−1,791	−1,335	−1,310
Anthracite	−218	−81	−22
Total net imports	−1,998	464	4,074
Total consumption	32,870	39,956	53,791
	(. . . . . per cent. . . . .)		
Net imports as per cent of consumption	−6.1% [b]	1.2%	7.6%

[a] Net exports denoted by (−). Excluding exchanges of electric power.

[b] 1947 happened to be a year of peak coal exports. Substituting 1948 exports would reduce this percentage to −4.3 per cent; substituting 1949 exports, to roughly −2 per cent. But whatever year was chosen, it would not affect the U.S. position as a net exporter in the early postwar era.

Source: Derived from data in U.S. Bureau of Mines, Minerals Yearbook, Vol. II, Fuels, various issues.

of coal exports, let alone raise them, or whether steam coal can return to a position of prominence is quite uncertain.

Largely through imports of crude and refined petroleum, the energy trade of the United States links it in varying degrees with other regions and nations (see Tables 8 and 9). The trade is dominated by crude oil and petroleum products imported from Canada, Venezuela, and the Netherlands Antilles. In 1965, about 80 per cent of the value of U.S. imports of energy minerals came from these countries. In that year imports of crude oil from these regions accounted for close to 70 per cent of the value of total crude oil imports. Venezuela remained the major crude oil supplier, but by 1965 Canada was supplying two-thirds as much, and its shipments to the United States were rising.

Imports of low-sulfur Nigerian and Libyan crude oil are a recent addition to the sources of foreign supply, but the volumes involved are as yet negligible. Imports of crude from the Middle East provide a small share as compared with those supplied from Venezuela and Canada. As for petroleum products—mostly residual fuel oil—Venezuela and the Antilles between them dominated the field.

Of U.S. exports of petroleum products, most go to Japan, Canada, and Mexico. These have been leading markets throughout the postwar period, although the United Kingdom (now a relatively small customer) figured prominently during the first postwar decade.

Nearly half of the country's coal exports in 1965 went to Europe, with about two-thirds of the balance going to Canada. These have been traditional markets, and little change has occurred in the last twenty years, except for the emergence of Asia as a consumer of significant amounts of American coal.

The Dimensions of the Energy Industries

When one considers the pervasive role of energy in the U.S. economy, it is perhaps surprising to discover that the share of its component industries in national income was less than 4 per cent in 1965 (Table 10). The energy industries are even less important, relatively speaking, as employers of manpower. Though one can argue over the definition of what types of business should be included in the energy category, a reasonable estimate is that the energy industries account for about 2 per cent of all manpower employed in non-agricultural establishments.

Energy bulks somewhat more heavily in consumer expenditures than in national income. About 7 cents of the consumer's dollar currently goes to direct energy purchases in one form or another (even this may strike some as low when compared with 5 cents per dollar that the consumer spends on alcohol and tobacco).

TABLE 8. Distribution of U.S. Foreign Trade in Selected [a] Mineral Energy Sources by Region,[b] 1965
(million dollars)

	North America	South America	Europe	Asia	Africa	Oceania	Communist countries	Total
Coal and other solid fuels								
Imports	13.4	–	1.4	77.3	–	–	0.2	15.1
Exports	157.4	21.8	229.2	77.3	0.1	0.2	8.0	494.0
Net imports	–144.0	–21.8	–227.8	–77.3	–0.1	–0.2	–7.8	–478.9
Crude petroleum								
Imports	335.2	507.6	0.7	293.8	50.4	–	–	1,187.7
Exports	1.5	–	2.3	1.1	–	–	–	4.9
Net imports	333.7	507.6	–1.6	292.7	50.4	–	–	1,182.8
Petroleum products								
Imports	416.8	431.7	3.9	10.8	0.6	–	0.4	864.1
Exports	99.2	33.0	108.5	128.6	22.2	14.8	2.3	408.5
Net imports	317.6	398.7	–104.6	–117.7	–21.6	–14.9	–1.9	455.6
Natural and manufactured gas [c]								
Imports	113.1	0.6	0.1	0.1	–	–	–	113.8
Exports	27.3	0.6	6.3	0.3	–	–	–	34.5
Net imports	85.8	–	–6.2	–0.2	–	–	–	79.4
Total, selected [a] mineral fuels								
Imports	878.6	939.9	6.1	304.7	51.0	–	0.6	2,180.8
Exports	285.4	55.4	346.3	207.2	22.3	15.0	10.3	941.9
Net imports	593.2	884.5	–340.3	97.5	28.7	–15.0	–9.7	1,238.9

[a] The table excludes electric power exchanges, trade in lubricants and a few other minor items. Total 1965 imports and exports of SITC code 3 ("Mineral fuels, lubricants, related materials") amounted to $2,222 million and $947 million, respectively.

[b] North America includes Trinidad and Netherlands Antilles; South America includes Cuba; and Communist countries (as defined by the U.S. Bureau of Mines) cover the European Communist countries (including Yugoslavia) and the Asiatic Communist countries.

[c] Includes liquid petroleum gas.

Source: U.S. Bureau of Mines, Minerals Yearbook, 1965, Vol. II, Fuels, p. 32.

TABLE 9.   U.S. Oil Imports, by Region and Country of Origin, Selected Years, 1950-1965
(million barrels)

Region and country of origin [a]	1950			1955			1960			1965		
	Total	Crude petroleum	Petroleum products [b]	Total	Crude petroleum	Petroleum products [b]	Total	Crude petroleum	Petroleum products [b]	Total	Crude petroleum	Petroleum products [b]
North America	119.4	10.1	109.3	139.2	22.7	116.5	185.0	43.9	141.1	333.5	110.3	223.2
Canada	–	–	–	17.1	16.4	0.7	44.1	40.9	3.2	118.0	107.8	10.2
Mexico	17.6	9.8	7.8	22.7	6.2	16.5	5.8	0.8	5.0	17.5	2.6	14.9
Netherlands Antilles	98.8	–	98.8	98.9	0.2	98.7	116.7	2.0	114.7	131.5	–	131.5
Trinidad and Tobago	3.0	0.2	2.8	0.5	–	0.5	18.0	0.2	17.8	48.2	–	48.2
South America	145.9	123.4	22.5	216.9	158.2	58.7	349.5	216.3	133.2	384.5	173.1	211.4
Colombia	15.7	15.7	–	8.1	8.1	–	15.5	15.5	–	18.5	15.2	3.3
Venezuela	130.2	107.7	22.5	207.5	148.8	58.7	333.6	200.5	133.1	363.0	157.9	205.1
Asia	42.7	40.4	2.3	113.8	113.2	0.6	150.9	139.2	11.7	155.4	144.1	11.3
Iran	0.1	0.1	–	3.1	3.1	–	12.4	11.2	1.2	29.1	28.6	0.5
Kuwait	26.3	26.2	0.1	56.3	56.3	–	66.6	60.4	6.2	22.1	20.2	1.9
Saudi Arabia	14.9	14.0	0.9	30.1	29.6	0.5	30.8	30.0	0.8	52.8	48.2	4.6
Indonesia	–	–	–	11.8	11.8	–	28.1	28.1	–	23.0	22.2	0.8
Africa	–	–	–	–	–	–	1.5	1.5	–	25.9	24.6	1.3
Libya	–	–	–	–	–	–	–	–	–	15.2	15.2	–
Total U.S imports [a]	308.6	173.9	134.7	470.0	294.1	175.9	687.2	400.8	286.4	900.7	452.0	448.7

[a] Regional totals include data for countries not shown separately, and the U.S. total includes negligible amounts for regions not shown separately. Shipments from U.S. territorial possessions are included in the total.

[b] "Petroleum products" includes small amounts of non-energy products, such as asphalt and wax.

Source: U.S. Bureau of Mines, Minerals Yearbook, Vol. II, Fuels, various issues.

TABLE 10. Some Dimensions of the Energy Industries,[a] 1947, 1955, 1965

	1947	1955	1965
A. Industrial origin of national income (billion current dollars)			
Energy extraction	3.4	4.2	4.1
Petroleum refining and related industries	2.5	4.8	5.1
Pipeline transportation	0.2	0.3	0.5
Utilities [b]	2.8	6.2	11.6
Total, energy-connected	8.9	15.5	21.3
Total national income	199.0	331.0	559.0
Energy as per cent of total	4.5%	4.7%	3.8%
B. Employment, persons engaged in production (thousands)			
Energy extraction	769	616	458
Petroleum refining and related industries	228	239	184
Pipeline transportation	28	26	19
Utilities [b]	502	599	638
Total, energy-connected	1,527	1,480	1,299
Total employment	57,705	64,221	71,248
Energy as per cent of total	2.6%	2.3%	1.8%
C. Expenditures for new plant and equipment (billion current dollars)			
Extraction, refining, pipeline transportation	2.16	3.44	4.39
Utilities (gas and electric)	1.72	3.65	6.72
Total, energy-connected	3.88	7.09	11.11
Total plant and equipment expenditures by business	24.15	32.25	57.32
Energy as per cent of total	16.1%	22.0%	19.4%

[a] The table suffers from lack of uniformity. For example, it omits rail, water, and truck transportation of energy materials, but includes pipelines and electricity transmission. Similarly, it includes utility distribution, while distribution of coal and oil is omitted. Results should, therefore, be considered as crude approximations.

[b] Includes, in addition to gas and electric, local sanitary services; latter represents minor component of total utility figure.

Sources: Panels A and B from U.S. Department of Commerce, The National Income and Product Accounts of the United States, 1929-1965. (Washington: Government Printing Office, 1966). Panel C from B. G. Hickman, Investment Demand and U.S. Economic Growth (Washington: Brookings Institution, 1965), pp. 232-33. Hickman's figures end with 1962. The 1965 figure is a rough estimate based on 1962-65 trends, as derived from other sources. Also, his data differ somewhat from other frequently cited investment expenditure data, and additionally do not incorporate the U.S. Department of Commerce revisions of the GNP accounts, issued in 1965.

In annual capital outlays on new plant and equipment, the energy industries are a leading segment of the economy. Their outlays account for about one-fifth of all business new plant and equipment expenditures. Compared with gross national investment of all kinds, the percentage falls in the range of 6 to 7 per cent.

In terms of the importance of the energy industries in international capital flows, the petroleum industry dominates. In 1965, it accounted for over 30 per cent of the value of U.S. direct investments made abroad, and it generated about 45 per cent of income on capital account from abroad.

One could go on and examine the importance of the energy industries in individual regions or states, their importance to relations with specific foreign countries, their relationship to depressed areas, such as in coal mining, etc. But the few benchmarks given above suffice to provide a useful perspective. Data needed to illuminate specific policy issues will be found in subsequent chapters.

Future Energy Requirements

For the scope of the present report, no purpose would be served by attempting to quantify the future demand for energy in the aggregate or for its different components except in the most general terms. Projections of consumption demands have been made by various agencies, public and private, and we shall have something to say about their uses and limitations. What is certain is that the demand for energy will continue to increase, that the future supply conditions for the separate sources of energy are too uncertain to permit confident predictions, and that many of the problems of energy policy will arise out of these uncertainties.

A review of recent estimates of future energy demands and of the resources available to meet them yields several conclusions: (1) Projections of demand for total energy tend to cluster in a range of 80 to 90 per cent increase from 1960 to 1980 and in the order of 200 per cent from 1960 to the year 2000. (2) There is a widespread belief that energy requirements for the next twenty years or more can be met without serious upward pressure on real costs. (3) It is commonly assumed that pressure of demand will affect adequacy first in the case of crude oil and natural gas, and that substitutes will be called for. (4) The prospects for growth in specific energy sources are subject to much greater differences of opinion than in total energy. (5) Nuclear energy, which is now a very small component of the total, is projected to increase at a rapid rate as a source of electric power. (6) There are sharp differences in the time pattern of future changes in energy mix and prices.

II

Oil

DURING THE LAST HALF CENTURY, petroleum liquids and natural gas have replaced coal as the main source of energy for the American economy. In 1920, coal supplied 78 per cent of the energy consumed; in 1940, 52 per cent; in 1965, 23 per cent. Oil and natural gas together rose from 18 per cent in 1920, to 44 per cent in 1940, to 73 per cent in 1965. In the petroleum constituent, natural gas has been overtaking oil. In 1920 consumption was negligible; in 1940 it was less than two-fifths the amount for oil; in 1965 it was nearly three-fourths as large as for oil. Though crude oil and much of the natural gas are closely associated in their physical origins and corporate ownership, they have become separate industries, different in market structure and in the public policies to which they are subject. We shall therefore assign them separate treatment, placing crude oil in the present chapter and natural gas in the one to follow. We have not specifically identified certain problems relating to natural gas liquids.

The Industrial Structure

Though oil is produced in thirty-three states, the major geographical source is the contiguous states of Kansas, New Mexico, Oklahoma, Texas, and Louisiana. In 1965 these five states produced 71 per cent of the national total. Texas and Louisiana produced 56 per cent of the national total, and Texas alone, 35 per cent.

The structure of the industry, from oil wells to consuming markets, is immensely complicated. The best vantage point from which to observe this structure is the refineries. To them flows the crude oil from the producing fields, mainly through a network of some 150,000 miles of pipelines, about equally divided between gathering lines and trunk lines. From them refined products flow out to marketing centers mainly by pipeline and water, and from there to local distributors mainly by truck. About one-half of the total refined products is in the form of gasoline or

24

other fuels for internal combustion motors. The remainder is mainly in the form of middle distillates and residual fuel oil used for boiler fuel in industry, for residential and commercial space heating, for diesel railway locomotives, and other uses. A minor fraction serves as the raw material for a wide range of products produced by the chemical industries.

In 1965, the twenty largest companies accounted for about 84 per cent of the domestic refining capacity. These same companies also own, either individually or in association, most of the pipelines. The pipelines are in effect technical adjuncts to the refineries, the avenues through which the companies convey the oil that they have produced or purchased in the field to the refineries, and through which they convey their products to marketing centers. The refining capacity is heavily concentrated geographically, most heavily on the Gulf Coast of Texas and Louisiana and in the Middle Atlantic states, but in lesser concentrations in the Chicago area and other North-Central industrial vicinities. California is a separate "province" where, in 1965, 84 per cent of the refining capacity was in the hands of seven companies. About two-thirds of East Coast oil is imported; the balance is received by tanker from the Gulf Coast. In addition to the refineries owned by the twenty largest companies, there are more or less a hundred smaller refining companies running from fairly large integrated companies down to quite small companies, geographically dispersed.

In contrast to the refining end of the industry, the business of producing crude oil is widely dispersed, though there is a tendency toward greater concentration. The twenty major refining companies [1] referred to above produce some 60 per cent of all domestic crude oil. The major companies are, however, integrated from the oil production to the refining stage in very different degrees. Some produce most of their oil; others are heavily dependent upon purchased oil. The remaining crude oil is produced by some 8,000 or more independent producers. A few of the larger "independents" are also vertically integrated through the refining and marketing stages. The independents run through a wide range of sizes, from large companies operating many wells in numerous fields to very small local producers.

Since most independents are engaged strictly in oil production, while the refineries and pipelines are mainly owned by the majors, they largely depend for their market upon the major refineries. The usual practice is for the oil to be bought at the wellhead by a major company on contracts

---

[1] Though it is part of the language of the oil industry, the distinction between "majors" and "independents" is extremely imprecise. The most relevant distinction is between "integrated" and "non-integrated," but even this cannot be regarded as very precise division because of degrees of integration. We use the top twenty merely as a convenient dividing line. For other purposes, the Chase Manhattan Bank list of thirty-odd companies is a convenient statistical point of reference.

running for an extended period. The contracts do not contain either firm amounts to be bought or firm prices. The amount to be produced by individual producers is regulated by state authorities, as we shall see later. The prices to be paid are "posted" in each producing field by the purchasing company; but as these postings are subject to change, contract prices cannot be firm.

In smaller fields there may be only one pipeline and, therefore, a single purchaser; larger fields are commonly served by two or more pipelines, so that there is some competition for the output. Shifts of producers from one purchaser to another are, however, exceptional because they involve a change of physical gathering connections and there is seldom any permanent price advantage to be gained. The pipelines are by law common carriers, and must transport oil without discrimination for producers who do not sell to the company owning the pipeline; but this constitutes a very small share of the pipeline traffic. Since companies are under no legal obligation to extend pipelines, some producers in new, poor, or remote reservoirs may be without pipeline connections and be forced to pay for truck transportation to pipeline, barge, railhead, or refinery.

Crude Oil Prices

The pricing policies of the major purchasers represent a rather complex field of inquiry that has not been subjected to close analysis in published studies. The major purchasers are large net buyers of crude oil from others, but they are also producers. Location and market factors affect price differentials among fields and grades of crude oil. Non-integrated producers desire the best price for their output. Integrated purchasers may want to buy at attractive prices; but they also have an interest in price stability for both crude oil and products, especially in the market for gasoline. Once a price structure is established, it has a strong tendency to perpetuate itself, except in belated response to a marked change in the market situation. Particular companies under- or over-committed for their supplies may at times contract for purchases at premium or discount prices temporarily without any overt change in the posted price in particular fields.

The principle of stability of posted prices does not, however, throw any light on why the prices for crude oil should not be higher or lower than they are. With respect to the oil that integrated companies produce for themselves, the posted price is purely nominal—an accounting entry in the intra-company records. If it is higher, more of the consolidated company profit appears to accrue at the producing level; if it is lower, more appears at the refining level—other things being equal. With respect to the oil they purchase, they gain by a lower price—again with the reservation, other things being equal. At lower prices, however, the competitive activity of independent refiners, or of some majors heavily dependent on purchased oil, might lower prices in the sensitive gasoline market.

A special feature operates strongly in the direction of higher, rather than lower, prices for crude oil, namely, the percentage depletion allowance for income tax purposes. This allowance permits a deduction of 27½ per cent from gross revenue from crude oil (with certain limitation) before calculating income tax. (See pp. 40-41). The higher the price of crude oil, the greater is this deduction. For companies producing a large fraction of their own oil, the net revenue advantages of a relatively high price are considerable.

Another line of thought may be introduced, relating to the interregional structure of prices. Prices for similar grades of oil are not everywhere the same; they may differ by reason of difference in location and transport costs or in accordance with varying local supply conditions. Nevertheless, there is a highly interconnected price structure. Posted prices are fully publicized, and there are market forces at work to keep them "in line." Publicity also operates in another way. The behavior of major companies is very visible; and if any of them appear to be discriminating in prices offered among regions, or offering prices notably below those of other companies elsewhere, they come in for severe public criticism from local producers; and they may be harried or penalized by the regulatory authorities of particular states. It is also worth notice that state governments have a direct interest in the price of crude oil, since production taxes are related to revenue and thus to prices.

The most important factor contributing to stable crude oil prices is not subject to control by the purchasers. The volume of crude oil production is regulated by state regulatory commissions; and this operates to dampen movements in crude oil prices. Given this foundation, which supports the product price level, the decision of the major integrated companies to take their apparent profits more at the producing level and less at the refining level may arise out of a variety of considerations, such as differential federal income tax treatment for production and the threat of price competition at the product level.

In spite of the masses of miscellaneous information on the petroleum industry—probably more than on any other American industry—little has been written or reported on the techniques or theory of the pricing of crude oil and of refined oil products. There is a good deal on gasoline prices because of the antitrust interest; very little on other products; very little on crude oil; and practically nothing on patterns of interrelationship between the crude and product levels. A study in this field would provide a degree of understanding of the economics of the oil industry that does not now exist, and would throw light on broader questions of interindustry competition. Without drawing up a full prospectus for such a study, or group of studies, some pertinent topics may be listed.

In the area of crude oil markets: (1) the character of price leadership in crude oil pricing; (2) the price effect of integrated companies buying from independents as well as from themselves; (3) the opportunities for price discrimination in purchasing crude oil; (4) the effect of joint venture crude pipelines and of crude oil exchanges among companies on the level and stability of prices; (5) differences between posted and realized prices; (6) the price stabilizing effect of state production controls and federal import controls; and (7) the functional relationship between the revenues obtainable from refined products and the price of crude.

In the area of product markets: (1) the price elasticities of demand of the various products; (2) the pricing practices of multiproduct refiners to maximize or "optimize" their profits; (3) the impact of brand name on product prices, especially in relation to gasoline price differentials.

The State Regulatory Systems

The oil-producing industry is subject to a complicated system of public regulation, unique in form and comprehensive in scope, which applies, however, only to the level of primary production of crude oil. It does not apply to the operations of oil companies at the levels of refining and product marketing. At these levels, the industry is on the same operating basis with the general run of American industries, subject only to the restraints of the antitrust laws.

Although each state runs its own system of regulation and the systems are not uniform, the large producing states have systems sufficiently similar that their characteristics can be described in fairly general terms. The activities arose out of the physical characteristics of oil production, property rights, and problems of market adjustment which were common to them all.

OIL RESERVOIRS AND THE "LAW OF CAPTURE"

Regulation of oil production grew out of an essential incompatibility between the nature of an oil reservoir and the character of the property laws that were applicable to producing oil from the reservoir. Oil in its original state is embedded in the interstices of porous rock surrounded by impermeable materials. It ordinarily exists under pressure from gas dissolved in the oil or contained in overlying gas-caps, or from underlying water—or from two or all three of these. When a well is drilled into the pool, oil is forced to the surface by the release of pressure. Since the pressures are common to the whole pool, the oil will flow to whatever openings are drilled into it. The essence of efficient oil engineering is that wells should be drilled at the points and in the number to secure the desired rate of flow, to minimize the loss of the activating pressures through escaping gas or water diversion, and to maximize the recovery of oil within economic limits. This means in effect that, in most cases, the efficient way to develop an oil reservoir is as a unit under a single operating authority. In the early days very little was known about the principles of reservoir engineering, and egregiously wasteful methods of production arose in part from this ignorance. But a potent and more permanent enemy of efficient production was, and still is, the law concerning property rights in mineral deposits.

Under the law in most countries, the minerals underlying the earth's surface are in the public domain, and do not belong to the surface owners. Under Anglo-American law, however, mineral rights go with surface ownership. This creates no particular difficulty in the case of solid minerals. The right is definable as applying to the specific minerals underlying the tract. But in the case of oil this is not so. When a well is drilled, oil will flow toward it; and it will drain not only the ownership tract in which it is located, but surrounding tracts as well, unless surrounding owners drill wells of their own. Moreover, the more wells any owner drills, the greater the share of the oil he can recover from the common pool underlying all the tracts.

In attempting to adapt the law of mineral rights to this unique situation, the courts arrived at the solution that the oil belonged to whoever brought it to the surface—the so-called "law of capture." This rule was based on an analogy of "fugacious" animals that roam the forest and plains across property lines. Capturing oil was like shooting a deer on your own property.

Applying this rule led to highly wasteful methods of developing oil reservoirs. Once a reservoir was discovered, tract owners around the discovery well had to drill wells of their own to protect their own interest. Wells drilled close to boundaries required adjoining owners to drill offsetting wells. Far more wells were thus drilled than were necessary for effective and economical drainage of the reservoir. There were numerous side effects. The motivating gas pressures were rapidly exhausted, leaving large amounts of the oil unrecoverable. Oil production often ran ahead of the facilities for handling and moving it, running into open pools, fouling streams, and creating fire hazards.

THE ORIGINS OF STATE REGULATION

State governments moved onto the scene in order to bring such chaotic and wasteful production situations under some control, under the general rubric of "conservation." They established regulations to limit the drilling of wells to tracts of specified size, to accomplish safe completion of wells, to regulate the uncontrolled escape of gas, to limit production from wells to amounts that could be currently marketed, and similar measures to create some order in the prevailing chaos in the field.

Another factor in the situation was access to markets. Independent producers were mainly dependent upon the large refining companies to purchase their oil and provide facilities—rail tank car, truck, or pipeline—for moving it. The purchasers were in a position to discriminate among producers as to purchases, and to favor their own oil where they were also primary producers. Early state regulation attempted to protect the producers by establishing the principles of equitable treatment of the producers in a reservoir.

At first, the regulatory activities were local in character: the establishment of rules to control the development of new reservoirs as they were discovered, one at a time, and to protect the interests of the producers in relation to their market. In time, however, the problems of marketing took on a wider significance. States would find that the producers in the state stood ready to produce much more than the purchasers wished to buy. In this situation, it was difficult to prevent the purchasers from discriminating among their sources and to assure all producers equal access to the market. Also, such a situation inherently contained a downward pressure on prices. In the end, therefore, the direction of regulatory activity was strongly toward limiting the total production within the state in order to support prices, and establishing formulas by which the statewide total could be divided as production quotas among all the producers in the state. By this route arose the institution of "proration," or the practice of "adjusting production to market demand." This procedure was first adopted by Oklahoma in 1929; it was soon adopted by Texas, and thereafter spread to the principal producing states east of the Rocky Mountains.

The change of emphasis in regulatory activity was especially activated, or reinforced, by the events of 1930-31, when great new fields in Oklahoma City and East Texas were repaidly developed. The oil seeking a market was so greatly increased that the price of oil was disastrously undermined. This event coincided with the onset of the Great Depression which created a marked weakness in the demand for oil. These two happenings concentrated the attention of the oil-producing states upon limiting the production and raising the price of oil. Control of production became, and still remains, the principal preoccupation of state regulatory authorities—shared, however, with a preoccupation with the rules of field development. It is to these two subjects that we shall devote the two following sections.

Responsive to the early problems described above, the early conservation statutes were written in terms of "prevention of waste." This phrase, however, is rather limiting for describing what regulatory agencies do. They fix the rules of field development with respect to well spacing and related aspects. They control production and allocate to individual producers with market stabilizing effects. What the regulatory agencies actually do is better understood if we forget the phrase "preventing waste" and look at their activities on the merits.

THE PRORATION SYSTEM

In the four or five years just prior to 1933, the states were mainly concerned to establish control over production and to firm the price of oil. To this end they strengthened their statutes and introduced new regulatory procedures. To many of those concerned, however, the problem seemed beyond control except by collective action among the states. To that end, a "Governors' Committee" was formed which recommended state production quotas and assisted the states in strengthening their regulatory organizations. The federal government also interested itself in the problem through the Federal Oil Conservation Board. Originally set up by President Coolidge after World War I to stimulate oil production at home and abroad in the face of an expected shortage, it came eventually to co-operate with the states in finding means to stem the tide of oil.

From 1933 to 1935 the industry was regulated by the federal government under the Petroleum Code of the National Recovery Administration (NRA). Among other things, the Code Authority assigned production quotas to the states, leaving it to state agencies to make them effective. When the NRA collapsed, the effort was resumed to achieve co-operation among the states to control production through an interstate compact; but this effort miscarried. An Interstate Oil Compact was in fact agreed upon and ratified by the Congress in 1935; but, due to disagreement among the states, its terms did not include the power to impose, or even to recommend, production quotas for individual states. The Interstate Oil Compact Commission has served as a forum for the discussion of mutual regulatory problems and has been the source of many reports through its technical committees; but it exercises no authority.

Failing a joint plan to regulate production, the several states have acted through their own separate agencies. There is therefore no direct control over the national total of production, and some states do not regulate production. But the methods by which the states exercise their separate powers have, in their combined effect, had a strong stabilizing effect upon the market. Foregoing a further historical account, we shall simply describe in outline how the system works, except for one comment. The excess producing capacity, which was the root of the problem in the 1930's, disappeared during World War II and for a few years afterward.

TABLE 11. 1965 Texas Yardstick, and Discovery and Marginal Allowables
(*barrels daily per well*)

Depth bracket (*feet*)	Yardstick						Discovery allowable	Marginal allowable
	10-Acre	20-Acre	40-Acre	80-Acre	160-Acre			
0 - 2,000	21	39	74	129	238		20	10
2,000 - 3,000	22	41	78	135	249		60	20
3,000 - 4,000	23	44	84	144	265		80	20
4,000 - 5,000	24	48	93	158	288		100	25
5,000 - 6,000	26	52	102	171	310		120	25
6,000 - 7,000	28	57	111	184	331		140	30
7,000 - 8,000	31	62	121	198	353		160	30
8,000 - 8,500	34	68	133	215	380		180	35
8,500 - 9,000	36	74	142	229	402		180	35
9,000 - 9,500	40	81	157	250	435		200	35
9,500 - 10,000	43	88	172	272	471		200	35
10,000 - 10,500	48	96	192	300	515		210	35
10,500 - 11,000	–	106	212	329	562		225	35
11,000 - 11,500	–	119	237	365	621		255	35
11,500 - 12,000	–	131	262	401	679		290	35
12,000 - 12,500	–	144	287	436	735		330	35
12,500 - 13,000	–	156	312	471	789		375	35
13,000 - 13,500	–	169	337	506	843		425	35
13,500 - 14,000	–	181	362	543	905		480	35
14,000 - 14,500	–	200	400	600	1,000		540	35

Source: Wallace F. Lovejoy and Paul T. Homan, *Economic Aspects of Oil Conservation Regulation* (Baltimore: The Johns Hopkins Press, for Resources for the Future, Inc., 1967), pp. 144-45.

In such a situation, when demand outruns capacity, the principal production control function is to hold reservoirs within limits set by the principle of Maximum Efficient Rate of Production (MER) in order to prevent premature exhaustion of the gas and water pressures. Beginning in the early 1950's, however, new producing capacity began to run well ahead of the increase in demand, especially in Texas and Louisiana. The cumulative increase of excess capacity placed an increasing strain upon the operation of the production control system.

The first step in the process of production control is the setting of a statewide total, usually done by the state regulatory commission one month in advance. The amount is determined by the commission on the basis of several types of evidence: a forecast of demand issued by the U.S. Bureau of Mines, the size and movement of stocks in storage, statements from major purchasers concerning the amounts they plan to buy (the so-called "nominations"), and a general assessment of the "feel" of the market. The going prices are taken for granted, and commissions do not think of their decisions as having a direct bearing on price, though of course they have a price-sustaining effect.

The next step in the process is to assign parts of the statewide total to individual producers as production quotas. This is done with the use of a formula under which (with exceptions to be noted) each well is assigned a "top" or "yardstick allowable." This hypothetical maximum is arrived at through a depth-acreage formula under which, at each depth, a well is assigned a yardstick allowable, which in turn depends upon the acreage of the tract on which the well is drilled. For purpose of illustration, we reproduce the Texas 1965 Yardstick in Table 11.

Taking the 40-acre column in the yardstick as a base, it will be seen that as the depth brackets go deeper, the allowable increases. Then, looking horizontally at each depth bracket, the size of the allowable increases as the size of the drilling tract increases, and decreases as the tract size decreases. Since the depth and acreage of each well is a matter of record, the yardstick allowables of all wells can be summed up into a statewide total. When the commission has decided what the statewide production is to be for the next period, this amount (less exemptions to be noted later) can be expressed as a percentage of the statewide sum of yardstick allowables. This percentage is known as the "market demand factor." Since each well owner knows what his own yardstick allowable is, he can simply apply the percentage to find his permissible rate of production for the next month. If his yardstick allowable is 100 barrels per day and the market demand factor is 30 per cent, he can produce at the rate of 30 barrels per day. It is an administratively simple, almost automatic, system which requires a minimum amount of policing because the major purchasers also know how much each producer is entitled to produce.

To guard against a possible misapprehension, a word must be said about what the proration formula does not do. The yardstick allowable of a well is in no sense a measure of its productive capacity. At the same depth bracket and in the same acreage bracket, a well capable of producing 100 barrels per day has the same allowable as one capable of producing 1,000 barrels per day. Thus, the statewide sum of well yardstick allowables is not a measure of statewide producing capacity. The fact that Texas is on a 30 per cent demand factor does not mean that it is producing at 30 per cent of capacity. The measurement of producing capacity lies in another sphere and has nothing to do with proration formulas, which are merely administrative devices. By the same token, no comparison can be made between the market-demand factors of different states. Each state has its own formula. If Oklahoma, Texas, and Louisiana all happened to be on a market-demand factor of 30 per cent, it would not mean either that wells in the same depth and acreage brackets were producing at the same rate, or that statewide production was operating at the same percentage of efficient producing capacity.

We turn now to exemptions from the proration process that we have been describing. Such exemptions are of several sorts. The most important is the exemption of so-called "marginal" wells—wells of low productive capacity that are permitted to produce to capacity. Marginality is defined by statute or by administrative rule. In Texas it is 10 barrels per day at shallow depth, moving by stages to 35 barrels per day at the deepest levels. In some states "discovery allowables" are allotted to the earliest wells in newly discovered fields for a limited period of time. In some states special allowables are assigned to secondary recovery projects. In Texas some whole fields with low average production per well are exempted. In the aggregate, these exemptions are by no means negligible. In Texas they apply to a major fraction of the wells in the state and to about 40 per cent of total production. The exempt quantities are deducted from the total for statewide production, and only the remainder is subject to the proration quota system described earlier.

An outstanding technical fact about proration is a systematic bias toward producing oil from the less productive wells. Production from flush, low-cost sources is kept severely in check. The limits placed on total production support the price structure under which the high-cost producers can continue to exist, instead of being forced into early abandonment.

This brief account of the proration system suggests certain lines of research designed to help clarify the effects of the system upon the production structure and the cost structure of the industry and upon the efficiency of its operations. The effects of proration in these respects are, however, intermingled with other aspects of regulatory practice to be noted in the next sections. Related studies will be discussed later.

THE REGULATION OF RESERVOIR DEVELOPMENT

The second main responsibility of the state regulatory agencies is to establish the rules under which oil reservoirs are developed. At the level of detail, this involves rules relating to efficient completion of well structures, preservation of gas pressures, avoidance of fire hazards, avoidance of stream pollution, and similar matters. Of primary importance, however, is the setting of well-drilling patterns for the reservoirs. This has an important bearing upon the degree of economic efficiency with which oil is produced. The subject is not one that permits a straightforward descriptive account, as in the case of proration systems, because the policies adopted through time have known many twistings and turnings. We shall therefore have to give a mixed historical and analytical account.

It is convenient to begin with two assertions. (1) In the usual case, development of a reservoir as a unit is the economically efficient method. This permits the number of wells to be kept to a minimum necessary for effective drainage, permits the ultimate recovery of oil to be maximized, and keeps the necessary capital investment and later production costs to a minimum. (2) This method of achieving economic efficiency has been almost universally violated in practice under the rules laid down by regulatory commissions, with great consequent additions to the cost of producing oil. We may first inquire why this has been so.

Looking back to the origins of regulation, we must recall the chaotic state of drilling under the "law of capture." Tract owners, or lease owners, drilled where they liked. As regulation was gradually imposed, it became the responsibility of the regulators to grant drilling permits and to specify the pattern on which wells could be drilled. In undertaking this task, regulatory agencies did not conceive their function to be the imposing of an efficient production structure; they were merely placing some restrictions upon the property rights of owners. This approach to reservoir development from the detailed side of well spacing, rather than from the side of efficient and economical development of the reservoir as a whole, has left its imprint on the producing structure of the industry.

Well-spacing rules consist essentially of specifying the minimum size of tracts upon which wells can be drilled and the minimum distance wells must be from the tract boundaries. The tendency has been for the drilling units to be increased in size, and 40 acres has come to be regarded as a basic standard. The early pressures for small tract drilling, however, caused much denser drilling to be permitted in most reservoirs. In the East Texas field, for example, something like 25,000 wells have been drilled, principally back in the 1930's, on an average spacing of around 5 acres; according to expert opinion, the number of wells is ten to fifteen times the number required for efficient drainage of the field. It has, moreover, been a common practice to permit at least one well to be drilled on any ownership tract, however small—a practice leading to "town-lot" drilling in settled communities. Where ownership tracts are typically large, it has been easier to establish relatively wide spacing; but even here there has been pressure for close drilling in order to hasten the recovery of oil under the proration formulas. Where drilling has been permitted on tracts of unequal size in the same reservoir, it presents the owners of small tracts with a license to drain the oil from adjoining tracts. Generally speaking, regulatory agencies now attempt to establish patterns of relatively wide spacing on drilling units of uniform size. But the heritage of the past clings to the industry in the form of the costs of a major fraction of total wells that were unnecessary for efficient drainage.

The relationship between well spacing and production quotas under proration is at once seen when we reflect that the proration depth-acreage formulas apply to all wells, except the exempt ones, no matter what the spacing pattern. The proration formulas determine how many development wells it is profitable to drill in an established field. The bias in most formulas has been toward encouraging over-drilling, because, for example, four wells on 40 acres would have a much larger combined allowable than one well on 40 acres. This bias has been partially removed in most current formulas by making the allowable proportional to acreage at any depth, at least up to 40 acres. There is no incentive to drill two or four wells to secure the same allowable as from one well. However, the older patterns of drilling have set their imprint upon the structure of production and the older proration formulas dominate the allocation of production quotas.

Related to the drilling of an excessive number of wells, but separate from it in principle, is the development of a capacity to produce greatly in excess of the market demand at going prices. It is estimated by the National Petroleum Council that nearly one-third of existing capacity is excess in this sense. The causes of this excess capacity are complex, being rooted in all the incentives that have existed for exploring for new reservoirs and for developing them. It is almost wholly confined to District III, which includes the major fraction of national production and most of the important market-demand prorationing states.

To some extent the incentives to overdrilling lie outside the field of state regulation, as for example the special federal tax provisions, to be noted later. But, given the discovery of new reservoirs, the incentives to overdevelop their producing capacity have largely arisen out of the proration practices and reservoir development rules described above. This is commonly criticized as one of the elements of inefficiency in the industry, leading to high costs. One the other hand, the excess capacity is also defended as contributing to national security by making it possible to increase production rapidly in times of emergency, such as the Suez Crisis of 1956-57 and the Arab-Israeli War of 1967. The emergence of excess capacity, it is clear, was not dictated by security considerations but was in the main an unintended by-product of regulatory practices. It is, nevertheless, a matter of interest to consider how far it may serve that purpose. This point should be kept in view in connection with the research proposals related to national security which are presented later (see page 42 ff.).

RESEARCH STUDIES RELATED TO STATE REGULATION

Assuming it to become an end of public policy to place crude oil production upon an efficient footing, studies in depth of the economic consequences of regulatory practices are needed to serve as the basis for constructive thinking about what changes might usefully be introduced. Such studies would be focussed around the two interlinking topics of production structure and cost structure. As adjuncts to a general consideration of energy policies, both lines of study would come eventually to focus upon the questions: How would the industry be best organized to assure its optimum contribution to future energy supplies? And what practical steps can be initiated in the direction of such organization?

1. Production Structure. The factual part of a structural study would best be built upon an analysis of actual reservoir development and proration practices, proceeding to statistical analysis of the well population. Such topics would be included as:

- *Proration formulas;*
- *Well spacing rules;*
- *Density of well spacing, by fields;*
- *Average production per well, by fields;*
- *Unused production capacity, by fields;*
- *Proved reserves, by fields;*
- *The statewide amount of production from unitized reservoirs (a) in primary production, (b) in secondary recovery projects;*
- *The statewide number of secondary recovery projects, and total production from them;*
- *Statewide well population, broken down by numbers in each of several acreage categories (under 5 acres, 5-10 acres, 10-20 acres, etc.);*
- *Statewide production, broken down into the same spacing categories.*

If, as will no doubt be the case, data on some of these and other topics cannot be obtained in the suggested detail, they still need to be kept in the forefront of attention as a guide to the kind of data that would be required for an accurate picture of the structural situation and for an understanding of the problems to be overcome if production is to be organized on an efficient cost basis.

On the normative side, analysis would be concerned with the means by which, from the factual state of affairs, progress could be made toward a more efficient production structure. Proration and reservoir development rules stand at the heart of this problem. Leaving aside obstacles rooted in vested and welfare interests, great ingenuity would be required to introduce innovations in the system that would, first, reduce costs in present fields; second, induce efficient organization in new fields; and third, provide incentives for the discovery of new fields. More will be said on this after we have looked into the impinging field of costs.

2. Cost Analysis and Supply Functions. The cost structure of the industry is intimately tied into the physical structure induced by regulatory practices. *Analysis of costs provides a necessary—indeed the most essential—part of the information required for policy purposes.* It not only brings into view the problems of efficient organization of the oil industry, but provides the basis for comparing the prospects for oil with other sources of energy in the total energy complex.

Apart from specialized studies of particular cost elements, such as drilling costs, the principal type of cost studies heretofore made within the industry is analysis of aggregate industry cost outlays distributed among functional categories. Studies of this sort in the crude oil industry have used the categories of finding, development, and production costs. These data are commonly used in an effort to arrive at the unit cost of a barrel of oil—a very tricky business that not uncommonly leads to misleading or invalid results. An element conducive to confused thinking is the fact that current cost outlays are heavily devoted to finding and developing reserves for future production, and do not reflect costs assignable to current output.

Within individual companies, such studies serve a useful purpose of a running accounting of investment outlays in relation to accumulation of new reserves and producing properties. When generalized, they are usually applied to quantifying a concept of "replacement cost." *Studies of this sort need to be subjected to critical review because of their proneness to misinterpretation, conceptual error, and invalid application to policy discussion.* Despite these problems, such studies may be of some use in the consideration of policy. The principal information they provide is a summation of current cost outlays, mainly investment directed to future production, laid alongside the reserves being currently proved or produced. *Further analytical study will need to be made of the types of cost information now being assembled by the industry. The uses of these data in the effort to determine the trend of unit costs need to be examined, and, in particular, attempts made to extrapolate unit finding costs into the future.*

The cost analysis peculiarly needed in relation to questions of energy policy is of a different kind. We state it here in terms of oil alone, but it runs across the whole range of energy sources as a means of establishing the economic marginal trade-off points between different sources. The basic piece of information is the current, or short-run, supply function, that is to say, the amounts of output that could be made available on cost-covering terms at different prices, in the absence of output restriction; or, to put it in another way, the marginal cost of producing successive increments of output. Though the data for computing such a curve would be highly imperfect, they should support a credible approximation. One result would be to quantify the difference between "economic" costs and actual costs, and raise the right questions concerning possible ways and means of minimizing the costs of oil supplies.

The next step is to calculate the increasing future cost of extracting larger and larger amounts from known reservoirs under existing technology. This would, of course, be much less precise than figuring the current cost schedule; but it could be revised as new information arises. Error-ridden data pertaining to costs in the future will get better as knowledge accumulates through time.

Research into the cost implications of new technology, though yielding even less accurate results, is still important. There are at any time new processes that appear close to fruition. These can be given due attention at any given time; and through time, evidence will accrue on their cost effects.

Beyond the effort to establish short- and long-run supply functions for known fields lies the highly speculative subject of the cost of finding new reserves, the amount of new reserves to be discovered, and the cost of developing them efficiently. Past experience may be extrapolated for a short period ahead to some useful effect, but beyond that there is little point in attempting cost calculations related to the completely unknown.

The final point to be emphasized is that cost analysis to support economically rational policy decisions, involving comparison between alternative sources of energy, has to be cast in the form of supply functions based on marginal cost analysis.

The Federal Involvement

Active regulation of the industry is confined to action by the states. The federal government is, however, involved in the operations of the industry, and to a degree in the regulatory process, through five lines of policy.

1. Under the Connally ("Hot Oil") Act of 1935, the federal government is committed to prevent the movement in interstate commerce of oil produced in violation of state law. The need for such action has largely disappeared and the machinery for enforcement has been discontinued.

2. The Interstate Commerce Commission has the power to regulate rates and service of interstate pipelines. Action under this power is no longer of much importance.

3. The federal government provides special tax treatment for petroleum as it does for other mineral extractive industries.

4. The federal government applies a quantitative restriction to the amount of crude oil and products that can be imported, in addition to a low tariff on crude oil and some products and a high tariff on the principal refined products.

5. The federal government is the owner of oil-bearing lands located in the public domain, including the offshore Outer Continental Shelf beyond the limit where state authority ends.

We will review briefly the last three of these.

DISTINCTIVE TAX PROVISIONS

All the mineral industries are subject to special provisions in the federal tax laws. Apart from certain types of capital gains transactions, the important special provisions applicable to the petroleum industry are of two sorts:

1. Intangible drilling expenses, which represent the major fraction of the capital costs of wells, are permitted to be charged as current expenses instead of being written off over the life of the wells. This has the same advantage, in extreme form, of any system of accelerated depreciation. It decreases current income taxes during a period of development and gives command over a volume of cash funds that may be thought of as interest-free deferred taxes.

2. The most important, and most controversial, special tax provision is the so-called "percentage depletion allowance." Under this provision 27½ per cent of the cross value of oil produced (up to a limit of 50 percent of net income) may be deducted before computing income for income tax purposes. In origin this was thought of as a substitute for an earlier method of allowing an annual depletion charge against the *capital value* of oil discovered, in lieu of depletion based upon costs incurred. Since, however, actual capital outlays are recoverable through current ex-

pensing of intangible costs and depreciation of tangible assets, the depletion allowance is in effect an extra benefit. Moreover, it goes on without limit of time or amount, as long as there is a net income. The theory supporting this provision is that it represents a method of recovering the *capital value* of a wasting asset, as opposed to the actual investment in finding and developing the asset. An additional argument offered is that the discovery of oil-bearing properties should be stimulated by tax remissions to those who are successful.

The distinctive tax provisions, in particular percentage depletion, are highly controversial and have been the subject of various studies or pronouncements—by industry agencies, Congressional committees, the Treasury Department, economists, and others. The grounds of disagreement run from the industry contention that they are necessary to the growth of the industry, at one pole, to the argument of some economists that they represent a misallocation of economic resources at the other, with various positions in between as to the equitable basis of tax discrimination. A dispassionate and comprehensive review of the whole range of past discussion on this subject would be a valuable contribution to knowledge.

Such a review, however, would not fill the gap in knowledge concerning the actual effects of the federal tax provisions upon the operations of the industry, and in particular upon the search for new oil reserves which is the primary point of interest for energy policy. It is the contention of the industry that elimination or reduction of percentage depletion would be detrimental to the growth of the industry. It would be highly desirable, if feasible, to establish some relation between tax revenue foregone and the new reserves discovered with the aid of the tax device. Unfortunately, this does not appear to be possible.

The primary barrier to such a study, so far as the large integrated companies are concerned, is that the additional funds provided by the depletion allowance are only one among many sources of funds that are available to oil companies to dispose of as they wish—for dividends, for development of existing reserves, for exploration for new reserves, for refining and marketing expansion, for investment in foreign operations, or for investment outside the industry. Insofar as these funds are in fact expended in the producing segment of the domestic industry, the further complicating factor arises that other incentives exist for investment in *development,* in order to obtain larger production quotas under proration *and* to secure larger tax benefits. The whole subject of the incentives for new exploration has to be explored among the interstices of this complex state of affairs. The specific effect of tax benefits on exploration probably cannot be isolated.

Any research into the relation of the distinctive tax provisions to invest-ment in exploration could be no more than a segment in much broader studies (1) of the sources and uses of investment funds within the in-dustry, and (2) of the incentives and disincentives to exploration con-tained within the existing structure and regulatory practices of the industry. These are subjects of critical interest for energy policy.

Within this broader context, it would be desirable for the results of the percentage depletion allowance to be, so far as possible, quantified in two ways: (1) the loss of revenue to the Treasury together with an analysis of the various economic consequences of reducing or withdrawing the special provisions; (2) an analysis of the cash benefits to various seg-ments of the industry (integrated and non-integrated, large and small, domestic and foreign operators) and to various classes of recipients (operating interests, royalty interests, and others).

In connection with the incentives for exploration provided by the tax provisions, several detailed questions arise which warrant study:

1. *How do the effective instrumentalities operate on business decisions? For example, is it the current expensing provisions or percentage de-pletion, or both, that induce outsiders in the high income brackets to take a share in wildcat drilling prospects? Is there a perverse effect of attracting investment funds away from exploration into development?*

2. *What is the rationale for extending these provisions to the foreign op-erations of oil companies?*

3. *Should the benefits of the tax provisions also accrue to the class of roy-alty owners?*

The special tax provisions for oil and gas raise questions over a number of aspects of public policy other than energy policy as such. For energy policy, their interest is as one of the complex of forces bearing upon fu-ture supplies and social costs of oil and gas.

IMPORT CONTROL AND NATIONAL SECURITY

Within a broad program of energy policy studies, one of the most important fields is that of national security. In the widest sense, the secu-rity of the nation is based on the strength and flexibility of its general economy. In this sense, all measures taken to assure the availability of a variety of substitutable energy sources is an element in security. As a subject for special study, however, national security considerations can be limited to kinds of precautions to be taken to assure energy supplies in a variety of contingencies that might arise in the international situation.

Not unnaturally, because of the predominant position of oil both in the general economy and in support of specifically military operations, energy for national security has been discussed mainly in terms of assured oil supplies. The specific policy most directly devoted to this purpose is the restriction of oil imports.

In 1959 the federal government put into effect a program of quantitative restriction of imports of crude oil and refined products. This action was the lineal successor to a report by a Cabinet Committee [2] in 1955 which stated that: "The committee believes that if the imports of crude and residual oils should exceed significantly the respective proportions that these imports of oils have to the production of domestic crude oil in 1954, the domestic fuels situation could be so impaired as to endanger the orderly industrial growth which assures the military and civilian supplies and reserves that are necessary to the national defense. There would be an inadequate incentive for exploration and the discovery of new sources of supply." Following this finding, the federal government authorities from 1955 to 1957 requested the large international companies to hold their imports down to the 1954 level. From 1957 to 1959, quotas were assigned to importing companies which they were under no legal obligation to respect. This voluntary plan having broken down by reason of the non-compliance of some companies, a mandatory program was initiated in 1959. The program was put into effect by Presidential Proclamation under authority conferred by the Trade Agreements Extension Act of 1955, renewed in 1958. The power was granted on the grounds of the requirements of "national security," the only grounds available to the United States at the time the action was taken for this departure from the general principles of the General Agreement on Tariffs and Trade (GATT). These principles, embodied in an international agreement, were designed to promote, not to restrict, the flow of international trade.

The facts behind the program were briefly as follows: Prior to 1948 the United States had been, at most times, a net exporter of oil, though this had in part been the result of restrictive policies toward imports during the 1930's. In 1948, imports of crude oil were at the rate of 270,000 barrels per day, or about 5 per cent of the rate of domestic production. Thereafter, imports rose to 700,000 barrels per day in 1954, and in 1958 to 1,200,000 barrels per day, or more than 18 per cent of the rate of domestic production. This increase was the result of the rapid postwar development of low-cost sources in the Middle East and Venezuela. In the meantime, after several years of rising domestic production, production declined after 1956. Producing capacity and the number of wells, on the other hand, kept rising, so that proration kept taking deeper and

[2] White House press release, February 26, 1955, on *Report on Energy Supplies and Resources Policy.*

deeper bites in the production rate of producers. There was every prospect that the proportion of imported oil would continue to rise, not merely cutting into the domestic level of production, but also undermining the price structure supported by state proration systems.

In these circumstances it could reasonably be argued that the incentives to find and develop domestic sources were being undermined. The supporting argument was that in the interests of national security, the domestic industry should be maintained on a basis of ability to meet any "national emergency."

The present line of policy is to limit imports of crude oil and products, except for the West Coast (and excluding residual fuel oil), in a fixed proportion to domestic production of crude oil and natural gas liquids (at present 12.2 per cent). It is left to experience to see how the industry performs under this degree of protection. On the West Coast, a deficit supply area, imports are adjusted to requirements and are proportionately higher. Restriction of imports of residual fuel oil has been virtually dropped, as having little bearing on stimulation of the oil-producing industry; though there do still arise questions of interruptibility of supply in time of emergency and the problem of interfuel competition with coal.

Imports are not allocated to particular source countries; importers are free to choose the source. There is no restriction on overland imports from Canada and Mexico, but the volume of exempted imports is deducted from the permissible total of imports to establish the permissible volume from elsewhere. Canadian imports have been rising, and this almost automatically reduces the volume from Venezuela, the main overseas source. Since one element recognized as of importance is protection of the interests of friendly source nations, this raises a troublesome sort of problem in foreign relations.

The close supervision over the total volume of imports is effected by allotting specific quotas to domestic refiners (about 127 in number in 1965) on the basis of their refinery runs of oil after allowance for certain historical quotas. The quotas are staggered in such a way as to allot higher proportionate quotas to smaller refiners, in line with the "small business" policy of the government. Interior refiners do not in fact process their quotas, since it is uneconomical to do so, but trade them to East Coast refiners in exchange for domestic oil—a transaction that is estimated to yield an average profit of about $1.25 on each barrel of quota traded. The essence of the quota system is, to a high degree, that the large international companies import and process oil they have produced in other countries, but pay for the privilege by buying the quotas of refiners who process only domestic oil. The system is a profit-sharing scheme which forces East Coast refiners to share the benefits of low-cost imports with inland refiners.

The total rationale of the import control program has never, so far as we are aware, been made the subject of serious analysis. Nor have many decisions modifying the program been supported by detailed explanation or justification. When it is stated that national security requires that oil supplies be assured, the statement needs to be clarified: "assured" to what extent, at what cost, by what means, and to meet what risks? And with what compromises among interested groups?

To the extent that national security is the basis for policies it is important to know the cost to the economy of this security. As economists have pointed out, to achieve absolute security might require devoting a large share of the nation's resources to security purposes, at the cost of a much lower rate of economic growth and standard of living. It is necessary to determine the trade-off points between higher cost and higher risks. But today scarcely even a framework exists within which to analyze this problem. Moreover, since there are alternative ways of achieving a security goal, the costs of the alternatives need to be analyzed if the national security cost is to be minimized. Cost-benefit analysis has been used in defense systems appraisal. Such an approach is not free from error, but it might well be adapted to studying the security ramifications of alternative energy policies.

The contribution of the oil import program to security was stated in terms of maintaining the incentives to explore for new reserves, thereby keeping the country more nearly self-sufficient and thus able to provide adequate supplies to meet undefined "emergencies." The cost to the public of maintaining the restriction was estimated in a 1962 government report [3] to be around $1.00 a barrel, or upwards of $3½ billion per year at present rates of consumption. *A levy of this size on the public suggests the propriety of an appraisal of the effectiveness of the program for its purported ends, as against the costs and benefits of alternative programs.*

Such an inquiry would also call for clarification of the security concept itself. A whole congeries of security concepts now invade the sphere of public discussion—a war contingency concept (for different kinds of wars), a national self-sufficiency concept, a Western Hemisphere self-sufficiency concept, a Western Alliance self-sufficiency concept, a good foreign relations concept, and finally, though without exhausting the list, a wholeheartedly protectionist concept that what is good for the domestic oil producers is good for national security. *These concepts swirl around in the controversial discussion of import policy, creating a general obfuscation of what is being talked about.*

In fact, in all discussions of oil import policy, much more is being talked about than national security. At issue is the entire structure of the

[3] Executive Office of the President, Office of Emergency Planning, *A Report to the President,* by the Petroleum Study Committee, September 4, 1962 (mimeo), p. 2.

industry. If import restrictions were to be abandoned, the present patterns of state regulatory practice would be deeply undermined and the high-cost segments of the industry swept away. If prices were much lower (say, by $1.00 per barrel), the domestic industry would gravitate into the hands of those capable of low-cost production. While a concomitant result would no doubt be a lower level of domestic production, it is by no means clear what the effect on the incentives to explore for new reserves would be in a system which permitted efficient low-cost enterprises to produce at much higher rates than those permitted under existing proration formulas. *Looked at apart from its security purpose, import restriction is a method by which the federal government assists the states in perpetuating a structure of production, costs, and prices imposed upon the industry by regulatory action. This being so, a study of import control policy will necessarily range over much wider territory than that defined by any national security concept. It will have to consider the more fundamental question of how the petroleum industry can be most effectively organized to play its part in the future energy complex. Conversely, the contribution of energy sources to national security will have to be analyzed in a wider context than that of oil alone.*

Given that the import control program reinforces the structure of the industry imposed by state regulation, can anything definite be said about its effects on the incentives to explore for new reserves? Here we can only repeat what was said earlier in connection with special tax provisions, that the operation of the incentives cannot be assigned to specific factors, but must be sought in a broad study of sources and uses of funds under the existing structure.

Within the security sphere itself, the import control program raises a number of special questions. To abandon it would no doubt force a crisis in the industry, leading to major reorganization. To reduce restriction gradually would impose pressures in the direction of reducing costs and securing domestic supplies from low-cost sources. The possible consequences of such a policy need to be studied, both in their security aspect and from the standpoint of productive efficiency and low-cost energy. In the same context, the possibilities of certain collateral policies need to be examined. One possibility, for example, would be to rely more heavily on imported oil for the present, at the same time directly subsidizing the exploration for oil and holding some of the discovered reserves in underground "stockpiles." Other proposals have been made which require study.

Within the immediate context of the present policy, there are questions concerning the implementation. Criticism, for example, has been widely expressed that the assignment of quotas to interior refiners serves no security purpose and is merely a bonus paid out of profits otherwise ac-

cruing to the great coastal refiners. Another view is that the quotas are necessary to protect the competitive market position of the interior refiners and through them to protect the position of the independent oil producers. An analytical study of the rationale and market effects of the quota system could well be made. In the course of such a study a fundamental question would be certain to arise: why any quotas should be assigned. Why instead should the government not recoup for the public benefit the value of the quotas by selling them at auction? Another question arises: Why have quantitative restrictions at all? Why not limit imports by the conventional protective device of the tariff? All these various questions would have to be taken up in a study of the implementation of a restrictive system.

The present import control program may be described as a holding operation. That is is to say, it supports the existing structure of production and regulatory practices and assures to domestic producers a major fraction of the expanding market. A study needs to be made of the program to determine whether this is the most effective and economical way to assure the optimal contribution of the domestic oil-producing industry to future energy supplies or to energy requirements for national security.

OIL FROM FEDERAL PUBLIC LANDS

The federal government is placed in a position of peculiar responsibility for future energy supplies by the fact that some of the most prolific potential sources are located in areas under federal ownership or control. This is especially true with respect to oil and gas in the Outer Continental Shelf and shale oil in the Western public lands. It is true in a degree of oil and gas in the other public lands and of the raw materials for nuclear processes. In the present chapter we shall be concerned only with crude oil.

Leases are issued by the Department of the Interior giving exploration and production rights on public lands and Indian lands. In unexplored areas they are issued to the first applicant at a low rental. When they are in the vicinity of known oil deposits they are sold to the highest competitive bidder. Leases of submerged lands on the Outer Continental Shelf are sold only on a competitive basis. Some 10 per cent of all domestic oil comes from public lands, about half of it offshore; and the amount is likely to increase since the Outer Continental Shelf off the Louisiana, Texas, and California coasts provides the most promising prospects for large new sources.

Because of single public ownership, development of reservoirs is free from those complications created by property rights in the states. The Geological Survey, which is the administrative body, has in general insisted on wide spacing and provides incentives to operators to adopt unit operation of reservoirs, but has not used its power to require unitization. In principle the federal government has the power to regulate production (or not to regulate it) in whatever way it wishes. In practice, it has made producers on the public domain subject to the rules of the states in which the lands are located. While there is no evidence of an intention to abandon this practice, at the instance of the Department of Justice, the Department of the Interior in 1967 made a public statement concerning its exclusive jurisdiction. The states have since then been urging that they be granted legal authority to include oil from the Outer Continental Shelf within their proration system.

With respect to the petroleum resources of the public lands, other than the Outer Continental Shelf, there appear to be good reasons for a critical review of the whole framework of present leasing policy. Questions have been raised concerning it that deserve answers: about how far it serves as a vehicle for speculative trading in leases; how far it encourages the postponement of exploratory effort; and how far it fails to require the most efficient methods of reservoir development. These and other questions relating to leasing policy will presumably come within the purview of the Public Land Law Review Commission, but such consideration needs the benefit of participation by persons with a broad interest in energy policies. Special questions relating to leasing policy for the Outer Continental Shelf also require study.

A more specific and potentially more far-reaching question can be raised concerning federal leasing and administrative practices on public lands including the Outer Continental Shelf, in relation to the regulatory practices of the states. A major deficiency of the oil-producing industry has been the inefficiency in the industry structure induced by state regulatory practices; in the future, rising costs and declining incentives to explore for new reserves may make improvement in this situation imperative. *The federal government is in a position where it could, if it so decided, introduce a code for oil production which would serve as a model for state action. The federal lands could provide a laboratory to see how an effective unitization program could be set up for new fields as well as old ones. In doing so, production on federal lands would have to be dissociated from the operation of state proration systems—to which it now conforms, not by legal necessity but at the option of the federal government. Present policy is to support the operation of the state systems. The alternative would be to demonstrate the methods by which they could be usefully modified. A preliminary study could bring out the problems to be encountered in such a change of policy.*

In another direction, a study could be made of the ways in which oil on public lands, including the Outer Continental Shelf, could be fitted into plans for national security. A problem of the future may be foreshadowed by the declining reserves-production ratio. The government could, by its own enterprise or by incentives to private enterprise, stimulate the discovery and partial development of fields, but hold them out of production. *Some part of the present reserves and a larger part of the new reserves could be kept under wraps and set aside as a national stockpile. The consideration of any such plan would have to be in the broader context of alternative measures for assuring energy supplies in the interest of national security.*

Incentives for Exploration

In a long-run perspective, it is a matter of great importance that the United States be assured of supplies of energy, from whatever source, to meet the needs of a growing economy. Since the present energy economy is so largely based on petroleum, it may be assumed that an expanding supply of oil will be a matter of concern for the calculable future. At present the sources of such expansion are heavily concentrated in foreign areas. Presumably, intense discovery effort can result in substantial additions to domestic supply, but at costs higher than for foreign oil. Given the present policy of restricting imports, the questions arise: How far will present policies induce exploratory activity, and how far will such activity result in new accretions of oil reserves? The answers will determine how far we may be driven to revise import policies or to rely upon substitute sources such as shale oil, or to add new incentives to exploration for oil. The policies now in effect may usefully be reviewed in relation to these fundamental questions.

The present incentives situation is, so to speak, the algebraic sum of different elements. Two federal policies, tax benefits and import restriction, are directed toward maintenance of relative self-sufficiency. While they undoubtedly contribute to this end, there is no evidence concerning the degree to which they do so. The force of the incentives provided by federal policies cannot be judged independently of the incentive structure contained within the state regulatory systems. These systems are not by design directed to the stimulation of exploratory activity. They are directed to market stabilization, adjustment of property rights, sharing of production among all producers, and waste avoidance in a limited sense. As we have seen, to a degree the state regulatory systems embody a pattern of disincentives for exploration.

Given a state of overcapacity, companies with large reserves may be disinclined to push the search for new reserves very hard. Since the proration system commits a large portion of the output to old high-cost wells, the rate of production from rich new deposits is severely curtailed. Given only modest expectations from exploratory effort, investment in development may be more attractive. Moreover, the more promising provinces for exploration, the offshore areas, involve very high expenses. In these circumstances, many companies are investing heavily abroad because foreign markets are growing rapidly, and perhaps also because they may expect that in time greater imports will come into the United States.

Many domestic producers argue that what they need is higher prices to provide more funds for exploration; but it is difficult to see how prices could be raised without even more severe restriction of production—not a promising route to stronger incentives to exploration. Moreover, there is no assurance that the added income would be spent on exploration. And the higher prices might be just what was needed to bring shale oil and liquefied coal over the economic horizon, reducing the market for crude oil. Finally, deliberately raised prices might cause those responsible for federal policy to think again about what level of imports was consistent with the requirements of national security.

The considerations just outlined demonstrate in what a clouded atmosphere the discussion of measures to stimulate the discovery of domestic reserves of oil now proceeds. The subject is central to any investigation of the problems of energy policy. It is not a topic that can be set aside for separate study; but from a broad general study of the oil industry, this aspect can be brought into focus as one of the most important topics to be investigated. In this broader study, the relationships among discovery, development, productive capacity, and deliverability are essential to the formulation of sound energy policies.

General Statistical Studies

The need, for policy purposes, of better statistical information on the industry has been highlighted in the *Petroleum Statistics Report* of March 1965, prepared by the Petroleum Statistics Study Group.[4] The topics to which they gave attention were Reserves, Productive Capacity, Wells Drilled, Transportation and Deliverability, and Expenditures and Revenues. (The *Report* also covered natural gas; and we may suggest in passing that a similar report on coal would be valuable.) *It would appear timely to have an early review of steps taken by government and industry agencies to implement the* Report. *Furthermore, the* Report *emphasized that statistical needs extended to refining, marketing, ownership, and international factors which lay beyond the limits at which the Group's work terminated—the delivery of crude oil to refineries. Another working group should take up this field of study.*

In the present context we shall say nothing further on statistical analysis in general, except to present a few thoughts on two of the topics, namely, reserves and productive capacity.

[4] Executive Office of the President, Bureau of the Budget, *Petroleum Statistics Report,* prepared by the Petroleum Statistics Study Group, March 22, 1965 (mimeo).

RESERVES

A standard system for estimating reserves is needed which goes well beyond the very limited scope of the "proved reserves" estimates prepared by the American Petroleum Institute. This provides only an "inventory" of known deposits in the immediate vicinity of existing wells under limiting economic and engineering assumptions. *The reasonable expectations of recovery from known fields are far greater and can be extended from the realm of relative certainty into degrees of increasing uncertainty. A system of categories of reserves, between which shifts are made over time, could produce highly informative results on the trend of amounts discovered in response to exploratory effort. (See also Chapter VIII.)*

It is further highly desirable that a well-based system of estimating "oil originally in place," by fields, be devised as a basis from which to establish trends in the degree of recovery of oil.

The sophisticated statistical treatment of reserves data is important not merely in estimating availability of oil within the calculable future. Reserves data are essential to statistical cost analysis, a field which can be highly informative, but can be, and sometimes is, highly misleading.

PRODUCTIVE CAPACITY

To quote the *Petroleum Statistics Report*, "Capacity to produce with existing facilities is one of the important factors which determine domestic petroleum availability. Relationships among production, productive capacity, and reserves are frequently used as indicators of the state of the industry and of its ability to meet requirements. Furthermore, productive capacity has a bearing on State regulatory systems."[5] It enters into judgments on the efficiency of the industry and on the conditions of national security. It is also of crucial importance for the cost analysis required to determine the conditions under which alternative sources of energy, especially shale oil, might encroach upon the markets for crude oil.

In spite of its importance, the art of capacity estimation is in a highly unsatisfactory condition. The two major industry sources of estimation (the Independent Petroleum Association of America and the National Petroleum Council) arrive at substantially different results; neither presents clearly defined concepts or statistical procedures. A third source (the American Petroleum Institute) has recently entered the field. Some relief from this increasingly confused situation is called for through the agency of a highly competent group assigned to critically review present methods and to propose standard concepts and procedures.

[5] *Ibid.,* p. 17.

III

Natural Gas

NATURAL GAS is a latecomer among the major sources of primary energy. As late as 1940 it accounted for only 11 per cent of total energy consumption, and provided only 36 per cent as much as crude oil. By 1965 it made up 30 per cent of the total and over 75 per cent as much as oil. Development of the natural gas industry was originally closely associated with crude oil; so far as the producing segment is concerned, this to a large extent is still the case. All oil reservoirs contain some gas, but proportions between oil and gas may vary greatly from one reservoir to another up to the point where gas is the principal or even the sole product. Until pipelines made long-distance transmission of gas feasible, natural gas had little economic value. Apart from some local use, it was largely important as an agent for flushing oil from oil wells. Flaring or venting dissipated much of it, thereby prematurely exhausting the natural pressure assisting oil recovery and leading to state conservation rules which limited the loss of gas in oil production. Where gas was found to have no association with oil, wells were plugged for lack of a market.

In the 1930's, the development of pipeline technology began to open up substantial markets for gas in distant urban areas. As a consequence, natural gas gradually replaced the uses of locally manufactured gas and also became an acceptable substitute for oil and coal for space heating and as an industrial fuel. As the market developed, the known gas fields were opened up, and the discovery of new gas fields became a special objective of exploration, even though the principal incentive for exploration still came from oil. In the development of gas markets, non-associated gas fields were preferred because they could be produced without regard to the effect on the production of oil. Where gas is associated with oil production, the amount of gas that can be produced is limited by state conservation regulations designed to protect oil reservoir pressures and by the amount of oil production permitted under the proration systems. It follows that, unlike non-associated gas, associated gas supplies cannot be adjusted to wide seasonal swings.

The Structure of the Industry

As might be expected, the geography of the natural gas industry coincides with that of the crude oil industry; but the regions accounting for the bulk of production are even more concentrated. The three most prolific regions are the Texas-Louisiana Gulf Coast, the Texas Panhandle and adjacent portions of western Oklahoma and Kansas, and the West Texas–East New Mexico region. The points of origin are relatively distant from the principal points of consumption in the North-Central, Northeastern, and West Coast regions.

Substantially, the structure of the producing industry runs parallel to that of the crude oil producing industry described in the preceding chapter. Since, until recently, most natural gas has been discovered in conjunction with exploration for oil, oil and gas are frequently produced jointly. But even where gas wells are not associated with oil production, they are still likely to fall within the same ownership structure as oil. For natural gas there is, therefore, a similar degree of concentration of corporate control in the hands of the major oil companies and a similar population of several thousand smaller producers. In the field of non-associated production there are, of course, numerous producers who are not in the business of oil production; some of these are pipeline companies.

While the corporate structure of the producing industry is to a large degree identical with that of the oil industry, the market structure is so completely different (together with the regulatory policies applied) that it must be described separately.

Natural gas in local markets superseded manufactured gas, mainly derived from coal, which was distributed by public utility companies subject to regulation of prices and service by state public utility commissions, on the same principle as electric utilities. During the 1920's, these companies had largely come under the control of the great holding company groups. Under the Holding Company Act of 1935, these groups were broken up and operations were restored to the control of the local operating companies. Moreover, local gas distribution and electricity distribution fell largely into the hands of separate companies, placing them in a competitive relation. The independent companies, seeking to introduce natural gas into their systems, had to look to some outside agency to transport the gas from the field to the "city gate." This function was undertaken by pipeline companies, some of which were affiliated with oil companies but most of them organized as independent entities.

The pipeline companies are mainly financed with debt financing, most of the bonds being bought by insurance companies on a basis of twenty-year amortization. A pipeline company enters into a long-term service contract with a local distributing company specifying volume to be delivered, usually for a twenty-year period. To assure sources of supply to cover the delivery obligation, the pipeline company enters into purchase contracts with field producers, specifying somewhat flexible rates of take for an agreed total over the life of the contract, commonly twenty years. The producing companies are required to make commitment for delivery from proved reserves. Thus, to contract for the delivery of, say, one billion cubic feet of gas per year for twenty years, the producer must have proved up at least 20 billion cubic feet of reserves at the time the contract is entered into—an arrangement which involves producers in a large amount of capital investment for development far in advance of delivery.

The production structure outlined above contains certain aspects which could well be the subject of research studies. Perhaps the most important of these would be a critical examination of the conventional contractual relations through the four stages of financing, distribution, pipeline delivery, and producer delivery to pipelines. The origins of the twenty-year feature are easily understood, for assurance to investors, assurance of delivery to communities, and assurance to pipelines of ability to meet their delivery commitments. Such long contractual commitments have been subject to criticism on the ground that they are no longer necessary and that they require large capital outlays for development long in advance of delivery requirements, exaggerating the costs passed on to the consumer.

An informational gap now exists as to the degree of concentration of ownership of reserves in the various areas. Another gap is the extent to which the known reserves are already committed under long-run contracts. These are matters which could have an important bearing on the performance of regulatory duties described below; but data are restricted by considerations of competitive secrecy among companies with respect to partially developed fields.

The relation of gas pipelines to primary producers and to consumers is very different from that of oil pipelines. In the case of oil, the refining companies (to the extent that they buy oil rather than produce it) buy it from producers at the wellhead, moving it to the refinery through pipelines which they themselves own, individually or in association with other refiners. From the refinery, refined products are moved to marketing centers through pipelines, mainly owned by the refiners. In the case of natural gas, the pipeline companies (except to the small extent that they own producing properties) buy gas at the wellhead, transport it on their own account, and sell it to distributing companies at the "city gate," or by direct sale to large industrial users. The great majority of consumers of gas are dependent upon a local monopoly for their supplies. Out of the special circumstances surrounding transport and distribution arose the special regulatory measures that are applied to natural gas.

Public Regulation of Natural Gas

Natural gas is subject to some form of public regulation at every stage of its progress from well to consumer. At the consumer end, retail rates and service are subject to state public utility regulation of rates and service. At the pipeline stage, wholesale rates and service are regulated by the Federal Power Commission, insofar as they are related to interstate commerce. At the producing end, field prices are regulated by the Federal Power Commission, again with the interstate commerce limitations. Also at the producing stage, natural gas is subject to some regulation under the same conservation statutes that apply to crude oil. At this stage, states do not regulate the field price of gas that is not covered by the FPC under the interstate commerce rule.

STATE AND LOCAL UTILITY REGULATION

The regulation of local rates and service by state public utility commissions stands somewhat apart from the subject of the present inquiry. It has, however the relevance that the retail prices fixed by regulation have a bearing upon the interindustry competition among different sources of energy. Something is said in Chapter V about public utility regulation as it bears on electricity; but for the most part we shall neglect the problems of regulating gas at the local distributive level.

STATE CONSERVATION REGULATION

The production of natural gas is to some extent regulated by state agencies under the same conservation statutes that apply to the production of crude oil, but because the market situations are different, the rules applying to gas are much less extensive. Where gas is associated with oil production, rules limit the amount of gas that can be produced in order to protect the reservoir pressures. The proration systems under which oil production is restricted and allocated to individual producers are not extended to gas. The rapid expansion of the market for natural gas has prevented any problem from arising with respect to the rate of output.

There are two principal kinds of state regulation. One consists of well spacing rules in fields where non-associated gas has been discovered. The specified drilling units are much larger than for oil, in some cases amounting to 640 acres, and seldom falling below 160 acres. The other kind takes the form of rules to protect the correlative rights of owners in common reservoirs. The rules require pipeline companies to take ratably from all the owners in order to give each of them equal access to the market—a device which runs into direct conflict with the terms of some old contracts and presents complications in arranging contracts for new supplies and proving up the reserves to support them. These difficulties are most easily surmounted when the owners agree to unit development of the reservoir, or at least place the contractual powers in the hands of a single agent.

State agencies may find a related cause for disagreement with pipeline companies when new pools are discovered in a large established field already served by pipelines under existing contracts. The owners of such pools naturally want to share in the market, but need some assurance of pipeline connections before going to the expense of proving up reserves; and state agencies are disposed to assist them by bringing pressure to bear on the pipeline company to establish connection and draw on the new source. While the law on the subject presents some obscurities, in the usual case the pipelines appear to be under no obligation to extend into new reservoirs; but they may be amenable to pressure and persuasion, especially since in an expanding market they will eventually want access. The problem is basically one of timing.

Generally speaking, the structure of the market for field gas is determined by the contractual relations between pipeline companies and producers under the rules and regulated prices specified by the Federal Power Commission; but the market structure is peripherally affected by the rules laid down by state agencies. In some circumstances, conflicts arise between state conservation regulation and FPC regulation of gas moving in interstate commerce.

FEDERAL REGULATION

The pipelines operating in interstate commerce (which carry most of the gas) are regulated as to wholesale prices (rates) and services by the Federal Power Commission under the Natural Gas Act of 1938. New pipelines, and extensions and abandonments of existing pipelines, have to be certified by the FPC on the basis of their gas supply, markets, project revenues, and costs. Federal regulation of sale price and service was introduced as an extension of the public utility regulation exercised by the states, and was designed to protect consumers from the monopoly powers of suppliers beyond the reach of state regulatory bodies. As in the usual public utility situation, the earnings of pipelines are limited to a specified rate of return on a technically defined capital value, usually in the region of 6 to 6.5 per cent.

During the early years of regulation, major decisions by the FPC established (1) the basis for original cost rate base, (2) a uniform system of accounts, (3) rate of return and cost allocation procedures. Rates were reduced and certification procedures were instituted. As is usual in public utility regulation, the Commission exercised close supervision over capital and operating costs as a basis for determining the net revenues applicable to cover the permissible rate of return. Although many pipelines produced a considerable part of their gas, a substantial cost item of the pipeline companies was the amount paid to producers for gas at the wellhead or gathering point. The price of the gas was fixed by free contract between producer and pipeline, and was taken by the FPC as a datum not subject to its control.

This situation was transformed by a decision of the Supreme Court in 1954. The duty of the Commission to regulate the field price of gas was not specifically included in the 1938 Act, and for a period of sixteen years the FPC assumed that it had no jurisdiction. Section 1(b) of the Act stated that:

> The provisions of this Act shall apply to the transportation of natural gas in interstate commerce, to the sale in interstate commerce of natural gas for resale . . . and to natural-gas companies engaged in such transportation or sale, *but shall not apply . . . to the production or gathering of natural gas.* (Italics supplied.)

In its decision in the Phillips case,[1] the Supreme Court ruled that this language required the FPC to take jurisdiction over field sales. The presumption that the first field sale was included within the concept of "production or gathering" was set aside in favor of including it within the concept of "sales in interstate commerce for resale." Production and gathering were limited in meaning to physical acts.

The influences which built up to this decision arose out of the rapid increase in the field price of natural gas as the market expanded after the war. The higher prices paid by the pipelines could be passed on in the wholesale prices paid by local distributors as legitimate costs, and by them to local consumers. The agitation for regulation of field prices, therefore, was raised mainly by municipalities or other representatives of consumer interests. Local distributing utilities played a part; while they could in principle pass the higher wholesale rates on in retail rates, this was dependent on action by public utility commissions, involving delay and loss of revenue. The pipeline companies stood somewhat aside from the controversy, being well situated to protect their own interests by passing on the higher costs. Producers were naturally violently opposed, and have waged a continuing battle to have the rule modified by judicial or legislative means.

The Phillips decision set the FPC a most difficult problem that has engrossed a large share of its attention ever since. It first attempted to apply the principles of pipeline rate-making, under which prices, where

---

[1] *Phillips Petroleum Co. v. Wisconsin et al.,* 347 U.S. 672 (1954).

the pipeline produced its own gas or a considerable part of it, are based on cost of service to establish a regulated rate of return on investment. This method, as applied to pipeline-owned production, has been controversial. The attempt to apply it to individual producers of natural gas, however, led into a morass of difficulties. If the price to each producer must be fixed in relation to his own ascertainable historical costs, a different price would have to be paid to each gas producer, even those in the same field selling gas to the same pipeline. Aside from the ambiguity of the resulting price structure, the administrative burden imposed by the necessity of determining a net revenue for each producer corresponding to a "fair rate of return" on his investment, proved intolerable. In addition, each producer was free to take legal action to support a contention that the price fixed for him was "confiscatory." These difficulties led the FPC to abandon the method.

AREA PRICE FIXING

Since 1960 the FPC has been working on a program to establish prices in particular areas based on some averaging of costs. The first decision of this sort came in 1965, in the case of Permian Basin in southwest Texas and southeast New Mexico. In the intervening years, the FPC, under direction of the Supreme Court (CATCO case),[2] kept the prices on new contracts "in line" with prevailing prices for like contracts in effect, thus halting the upward trend of contract prices; and this situation continues with respect to prices in other fields in which rate investigations are under way.

A troublesome problem for the FPC in exercising control over field prices of natural gas is that the two ends it seeks to serve—the protection of consumer interests and the stimulation of the search for new sources of natural gas—may at times be in conflict. Moreover, the application of the principle of basing prices on average cost may work without plan or design in one direction or the other. The Supreme Court decision in the Phillips case was heavily weighted with the idea that the 1938 Act was designed primarily to protect consumer interests and that this required carrying price control back to the source. But the rapidly rising requirements for energy in general tend to shift emphasis to enlarging or replenishing the sources. In its Permian Basin decision[3] the FPC attempted to take supply into account by establishing one level of prices for gas from

[2] *The Atlantic Refining Co. et al.* v. *Public Service Commission of New York et al.,* 360 U.S. 378 (1959).

[3] Federal Power Commission, Opinion No. 468, Docket Ar 61-1, G-18466, Area Rate Proceeding, August 5, 1965. The decision runs to 139 pages. Federal Power Commission, Release No. 13964/J-7866, August 5, 1965, announces the decision and contains the Commission's own summary.

"old" sources, and a somewhat higher level from "new" sources to have an incentive effect on exploration. The two levels were based on different types of cost evidence: the first upon evidence of historical costs in the Permian Basin, including an allocation of costs incurred jointly with oil exploration and production; the other, upon a nationwide sampling of current costs incurred in developing new gas supplies, where production was not associated with oil production.

The first field of research that suggests itself is that of the methods of cost estimation used for establishing field prices. The effort to base field prices on average costs, arrived at by either of the methods mentioned, leads into troublesome technical problems of cost estimation. We shall not attempt to describe the difficulties. To a considerable degree they arise out of the intimate association of oil and gas costs. The problem is not simply one of dissociating gas costs from a well-established joint total; it carries back into the extremely limited state of knowledge concerning the costs of finding, developing, and producing petroleum properties. In the case of the cost estimates for non-associated gas, used as the basis for gas from "new" sources, an additional set of problems arises. The cost analysis by the FPC is admittedly experimental. When such far-reaching consequences are built upon such tenuous methods, it can hardly be questioned that the methods of cost analysis utilized for price fixing should be subject to expert appraisal.

A subject for critical study, of more fundamental importance than the technical cost formulas themselves, is the rationale for using such cost analysis as the basis for fixing prices. The compulsions on the FPC are plain enough. Though the point is arguable, the Supreme Court appears to have compelled the Commission to fix prices associated with costs for the protection of consumers. It devised types of cost study to fulfill this obligation, and sweetened the results of one of them in the Permian Basin decision to provide an "incentive price." What needs to be investigated is the nature of the connection, if any, between this type of average cost formula and the supply schedule for natural gas.

As we saw in the case of crude oil, a supply schedule is built up from the cost of marginal increments of supply. Data to support this type of analysis is what is most needed to establish relationships between gas prices and supplies. Because of the intimate physical associations between oil and gas supply conditions, especially in the exploratory stage, some part of the data will necessarily arise out of what are primarily oil cost studies. However, in view of the extent to which cost factors bearing on gas supply may be separable from oil costs, especially in connection with directional drilling in gas provinces, something in the nature of a useful, if highly imperfect, gas supply schedule could possibly be built up, stage by stage. At present, however, statistical information is very de-

ficient with respect to exploratory drilling distinctly related to natural gas.

Knowledge of this sort would be useful whether or not prices were regulated. If they were regulated, it would, first, provide the regulators with a better basis of judgment concerning the consequences of their acts; and, second, provide those responsible for a general energy policy with a basis of marginal cost comparison with other energy sources. If they were not regulated, it would still serve the second of these purposes.

SHOULD FIELD PRICES BE REGULATED?

While the field prices of natural gas are in fact being regulated, the argument goes on as to whether this procedure makes sense. The economic and social rationale for field price control is by no means firmly established. Regulation applies to a situation without parallel or precedent in American regulatory experience. Leaving aside the views of obviously self-interested parties, we will note briefly the controversial points that arise in serious economic discussion.

The strongest argument brought against the procedure is that price control is unnecessary because the market for natural gas in the field is sufficiently competitive. Some economists have come to this conclusion in their studies. Other economists have arrived at a somewhat different conclusion, not so much because of monopoly power on the sellers' side, but because of the peculiar characteristics of purchase contracts and of the market structure.

One special feature of such contracts, until disallowed in recent years by the FPC, has been the "most-favored-nation" clause under which the sellers under old contracts received the highest price later paid by the purchaser to any seller. The other special feature is that contracts are for delivery over twenty years from reserves already proved. A part of the argument of those favoring regulation is that buying under such contracts at times of rapidly expanding demand, by tying up vast reserves may create shortages of reserves available for purchase, exerting strong upward pressures on prices. The sum of these two features has the effect of creating enormous economic rents or "unearned increment" in the hands of owners of producing properties. The heart of the argument of consumer interest is that these rents should be shared, rather than retained wholly in the possession of producers.

One point advanced by the economists who find the industry "workably competitive" is that in the early stages of development of the market a considerable share of monopsony power rested in the hands of the pipeline purchasers. With usually only a single pipeline serving a field, and with relatively large supplies available, the pipelines were able to contract for supplies on very favorable price terms. The rapidly rising prices after World War II resulted from the expanding demand which led companies to extend multiple pipelines into the large fields and to compete with one another for the available reserves. In this view, with numerous producers in the field there was always competition on the sellers' side and increasing competition on the buyers' side. Rising prices were the result of competition in an expanding market.

Conceding a part of this analysis, those taking the position that price regulation is justified have relied mainly on the arguments: (1) that the contracting for large blocks of reserves created at least temporary shortages of reserves offered for sale; (2) that the favored-nation clauses created unjustified gains under old contracts; (3) that the ownership of a large part of the richest reserves was highly concentrated in the hands of a few large oil companies; and (4) that these facts placed an undue amount of bargaining power with respect to price in the hands of the producers. In its Permian Basin decision of 1965, which fixed prices for that region, the FPC committed itself to that argument, with one member dissenting.

The most important field of economic investigation is concerned with this fundamental question: the justification of any regulation of field prices. Present evidence demonstrates that differences of view will persist, even among disinterested observers. But the whole field of argument needs to be objectively reviewed. It is not yet at all clear what lines of policy are most appropriate for fitting natural gas into a broad program of energy policy.

The nature of the market for natural gas in the field is so unique that it does not correspond to the simplified conventional views on policy, which divide markets into two sorts: competitive markets subject to the antitrust law policy, and monopolistic markets subject to public price regulation. The field is, therefore, left wide open to arguments between interested parties and between objective observers, without any clear guidelines to policy. On the political front, it is not clear that the Congress ever intended the Natural Gas Act to require regulation of field prices; the Supreme Court decided that it did, or at least that the wording of the Act required it. The Congress has twice voted to free field prices from regulation, but the bills were vetoed by two Presidents. *Within this mixed political and judicial situation, it is important that the issues of policy be clearly articulated. Within the historical context, there should be no initial presumption that the actual procedures adopted by the FPC have any rationale other than that of attempting to fulfill an obligation imposed upon it by a decision of the Supreme Court. The criteria formulated by the Court became an important part of the investigation. Other criteria have to be examined. The effects that field price control may have for the role of natural gas within the total energy complex have to be appraised.*

Pipeline Co-ordination and Planning

With respect to the pipeline transmission network, a study could usefully be made of the means of improving the cost-efficiency of the system. Heretofore, pipeline certification has proceeded on a case-by-case approach, which leads to duplication and cross-hauls. A regional approach to certification needs to be examined, to determine feasibility and possible economies to be derived from pooling or co-ordinated use of pipeline facilities and from jointly owned facilities. Possibly the Outer Continental Shelf offers a special field for advance planning in this respect.

Interindustry Competition

Once an advanced technology for pipeline transmission had been developed, the rapid penetration of natural gas into the markets of coal and fuel oil was based on the relatively low field prices upon which supplies were available. At first the competition was mainly for the space heating market, in which oil was replacing coal. Gas hastened the disappearance of coal and began to replace oil or, rather, limit the expansion of its use. Even where natural gas did not have a distinct price advantage, its superior cleanliness and convenience supported its acceptability. Later, gas penetrated deeply into the industrial fuel field in direct competition with coal as a boiler fuel. The degree of this penetration was partly based on geographical factors: gas was most widely used nearest the source; coal retained advantages near its source. Eventually, special cost factors, to be noted later, enabled gas to penetrate more heavily into the markets of coal.

The competitive relationship between oil and gas is marked by the peculiarity that both are products of the "petroleum" industry and that natural gas is mainly produced by "oil companies." There is the further relevant fact that at the wellhead oil is the more valuable product in terms of price per Btu energy content. This is mainly because of the greater unit cost of transporting gas in Btu terms and the inability of gas to compete directly with gasoline or other oil products that are more valuable than ordinary fuels. In these circumstances, it might be thought that the oil companies would have kept something of a lid on the production of gas in order to relieve oil markets from the competition of gas. But the competitive structure of the oil industry did not lend itself to this solution. As pipeline demand developed, individual producers were glad to receive the extra income from this product with limited value locally.

The ultimate inroads of gas into the markets of refined oil products, especially space heating, were not subject to any collective control by the oil industry.

The competitive entry of gas into uses mainly served by coal, especially as industrial boiler fuel, was assisted by the economics of pipeline operations. Two factors were at work. In the first place, the local demand for gas at retail is highly fluctuating and seasonal because so heavily dependent on space heating. To meet this situation, pipeline companies adopted the device of the "interruptible" contract. Under such contracts gas is sold directly to industrial users, at prices not subject to FPC control (except as to some allocation of fixed charges, as indicated below), on condition that the service may be cut off when the gas is needed to supply local consumers. The low price available under such contracts may make it worthwhile for industrial users temporarily to suspend the use of coal, especially in connection with the operation of electrical generating stations, though there appears to be little information on where and to what extent "interruptible" gas is actually interrupted.

In addition to seasonality, the other factor affecting competition with coal arises out of the economics of large-diameter pipelines. The unit cost of transportation declines rapidly as the diameter of the pipe increases. In planning the facilities to serve local communities, the pipeline company may build much larger lines than are required for that purpose, possibly contracting for the sale of the surplus to industrial users at prices lower than those fixed by the FPC on wholesale sales to local distributors. In the usual circumstances, rates to local consumers are probably lower since, with smaller pipelines, the charges would be higher. Given large pipelines, however, a problem arises in that the lower the contract price to industrial users, the less they contribute to the fixed charges of the pipeline system as a whole. The FPC does not have direct jurisdiction over industrial contract sales; but by formula covering the distribution of fixed charges between wholesale and private contract sales, it can assure that industrial users contribute to meeting fixed charges to some determined degree.

Both these practices—interruptible sales and firm sales at relatively low prices—are naturally not well regarded by the coal industry. The damage to the industry comes mainly in the one large market where coal has retained a strong competitive foothold, as boiler fuel for electrical generation. The subject is one that deserves an objective economic study, with special reference to (1) the economic desirability of interruptible contracts and possible alternative solutions to the problem of seasonality, and (2) the issues connected with extending the use of gas in industrial uses through differentially lower prices.

To the extent that gas prices are kept low by regulatory rules, they en-

courage the wider use of gas. Apart from its competitive aspects, this raises a "conservation" issue. Should the using up of the uncertain reserves of natural gas be deliberately hastened by regulatory action? The point that the use of natural gas as industrial boiler fuel should be regarded as an "inferior use" and discouraged has been especially urged by the coal industry. But the issue thus raised has much broader implications: In what ways, and for what reasons, should an energy policy be concerned with problems of end use of the various energy sources? This problem deserves special study.

Related to the end-use problem, a new factor, air pollution control, is entering the potential competitive position of natural gas. If certain types of solid and liquid fuels are excluded from use in urban areas unless they have undergone costly purification processes, natural gas might be elevated to the position of a "superior" rather than "inferior" fuel for boiler fuel use. The ramifications of pollution control for energy policy will be taken up in the final chapter of this report.

The role of gas in the future energy economy may be greatly affected by factors that have not yet achieved practical importance. One of them is the possibility that large imports of natural gas may be available, both overland from Canada by pipeline and in the form of liquefied natural gas (LNG) from overseas. The Federal Power Commission has authority over imports and exports, but no general policy exists. Close study of the way in which imports may be incorporated into the supply structure is called for as a basis for policy, first, in the special Canadian case and, second, in the general overseas case. The Canadian case is already building up a sensitive issue about the way in which U.S. regulatory action may include an attempt to influence the price at which gas is sold in Canada. The considerations bearing upon import policy are much wider than those within the specific sphere of responsibility of the FPC. They will nevertheless become intermingled.

Another factor bearing upon the gas component in the energy "mix" is the possibility that synthetic gas from coal may become economically feasible. At present the principal subject requiring study is the amount of R&D effort that should be mounted to demonstrate the possibilities. In the outcome, however, the results might force reconsideration of the basis of policy with respect to natural gas.

General Statistical Studies

At the end of the preceding chapter on crude oil, we incorporated by reference proposals of the Petroleum Statistics Study Group for a program of statistical documentation and analysis covering Reserves, Productive Capacity, Wells Drilled, Transportation and Deliverability, and

Expenditures and Revenues. The proposals to a large degree also referred to natural gas, and are here also incorporated by reference. (See page 50.)

In this connection, we wish again to emphasize the importance of further study of methods of estimating reserves. As a general problem for all the energy industries, this is given special attention in Chapter VIII.

IV

Coal

COAL PROVIDED THE ENERGY BASE for the industrialization of the American economy in the last half of the ninteenth century and the early part of the twentieth century. It underpinned the growth of the steel industry and the structure of industries based upon it; it fueled railways and ships; it drove power generators; and it took over the domestic and commercial space heating field.

In 1850, over 90 per cent of the fuel used was wood; coal contributed less than 10 per cent. By 1900, coal (including both bituminous and anthracite) could claim over 70 per cent, wood 21 per cent, and oil and natural gas jointly 5 per cent.[1] As late as 1940 coal still contributed about 50 per cent. But by 1965, coal's share was down to less than 25 per cent, with oil and gas up to more than 70 per cent.

The decline of coal in favor of petroleum was not merely relative; it was also absolute. The consumption of coal in 1965 was less than in 1910, though the absolute peak was just after World War II. During the nineteenth century, anthracite provided a substantial fraction of coal supply; but it has faded away to such a negligible proportion that matters relating to coal today may be discussed solely in terms of bituminous coal.

Though reduced in relative importance, the coal deposits of the United States remain one of our most important potential industrial assets. As presently estimated, resources and reserves of coal, however defined, are very much larger than those for either oil or gas. But that is one of the few statements about future coal supplies that can be made with confidence, for such estimates are highly speculative. Depending upon assumptions as to depth, thickness of seams, methods of recovery, and statistical probabilities, they display a wide range of possibilities. This is a major question that needs looking into. We shall return to it later.

---

[1] These figures differ from those in Table 2. That table excludes wood because of its unimportance for most of the period covered.

The Structure of the Industry

Four main factors explain much of the turbulent history of the U.S. coal industry: (1) easy entry in the early days due to low capital requirements with consequent predominance of small mines and vulnerability to business fluctuations; (2) high incidence of transportation cost and importance of federal railroad legislation to regulate rates; (3) highly immobile labor force with tendency toward high levels of unemployment, due to lack of employment alternatives in coal-mining regions, and consequently low wages; (4) technical backwardness and lack of research and development efforts on the part of industry. In the past two decades much has changed, but before dealing with recent developments certain structural features of the industry will be outlined.

The principal highly developed coal fields are in three regions: the Northern field based in Pennsylvania and northern West Virginia; the Southern field based in southern West Virginia and Virginia; and the North Central field based in Illinois and Kentucky. About 75 per cent of the coal now produced is from these states. There are potential reserves of substantial volume farther west, mainly in the Rocky Mountain region. These have in the past been only slightly developed for want of supporting markets, but are now coming into use owing to the economies of long-distance electricity transmission combined with cheap surface-mining methods. There are also large amounts of sub-bituminous coal and lignite scattered through northwestern areas, but they are still outside the competitive economic margin.

While coal sources are in, or near, the most highly industrialized areas of the country, this advantage is offset by the high cost of transporting coal, as compared with pipeline transmission of crude oil, oil products, and natural gas. Appalachian coal cannot penetrate competitively very far toward the southwestern oil and gas producing regions. Oil and especially gas can penetrate into coal market areas as competitive fuels.

Railroads represent the main means of coal transportation. Therefore, the marketing of coal is very much affected by the level and structure of freight rates. A brief description of the history of rail rates suggests reasons why until recently not much effort was directed toward modernizing the transportation of coal.

Before the advent of the Interstate Commerce Act of 1887, the practice of quoting coal prices on the basis of delivered prices had become general. This was primarily a device of the railroads to secure traffic from mines to markets. On this basis competition among mines for markets was intense, as well as the competition of railroads for traffic. Minehead price realization from sales made a chaotic pattern, as each sale yielded a different net of delivered price less transport costs. The pricing system was made more complicated, and discriminatory, by the fact that the railroads themselves became large proprietors of coal-producing properties. This was especially true in the anthracite regions, but to a degree also for bituminous coal. The railroads could discriminate as to rail services in favor of their own coal, thus wielding a power of life or death over many independent mine owners. The railroads were also capable of discrimination based on the competition among themselves. In markets where there was rail competition, prices might be cut to the bone, while advantage was taken of the monopoly power in markets served by a single line.

The peculiar relationship of the railroads to coal production and marketing was one of the strongest factors behind the passage of the Interstate Commerce Act. Under the "commodity clause" of this Act, the railroads were prohibited from themselves engaging in coal mining and transporting their own coal. In other ways the Interstate Commerce Commission gradually rid the industry of the grosser forms of discrimination. On the other hand, the rate policies of the ICC followed practices which had grown up in the railroad industry. It kept the basic rate zones within which rates to destination were uniform regardless of distance. And it stood ready to approve special rates to permit producers to sell in distant markets. This heritage from the past was accepted by the ICC because it did not wish to disrupt the market structure that had been built up. Railroad rate structures were further affected by the existence of water transport in the Monongahela, Ohio, and Mississippi river systems which served wide producing and market areas. Water transport being cheap, railroad rate structures had to be adjusted to this competition.

In these conditions, local markets had a wide range of sources upon which to draw; and in slack periods the competition for these markets became severe. Since the coal industry was so dependent on industrial demand, it was highly vulnerable to the influence of business cycles. In consequence, periods of intense cutthroat competition were common, prices were volatile, and the fortunes of most producers were precarious. Competitive behavior in the industry was much affected by the fact that it was mainly made up of thousands of small producers. Only a few relatively large companies grew up. This structure was generated by physical features of the sources. Much of the coal could be mined from outcroppings or at shallow depths. Entry was easy with relatively small capital investment. The principal costs of operation were labor costs. Out of this fact rose a chronic state of excess producing capacity. When markets were strong, local groups would get into the business. When they became weak, the capacity did not disappear. This intensified the competitive struggle.

Another characteristic of the industry was that it came to be chronically overmanned with an excessive labor force. At the best of times production was highly seasonal, so that the incidence of seasonal employment was high. But beyond this, a large degree of chronic unemployment was endemic. Mining mainly took place in isolated communities, lacking alternative sources of employment and with a relatively immobile population. This fact tempted owners to attempt to translate some of their own market difficulties into low wages, an expedient which naturally intensified price competition. Effective union organization was difficult because of physical isolation, the diverse national sources of the labor supply from recent immigration, and the active opposition of owners, as well as because of excess labor supply. Very tense and disturbed labor relations resulted. When relatively effective union organization was achieved in some areas, especially in the Northern field, it resulted in competitive loss of markets to unorganized regions.

Another consequence of the small-scale character of mine operations was that mining was carried on with extreme technical inefficiency. There was a minimum of research and development of advanced technology, and small owners usually could not afford to install such as there was. Owners found wage-cutting a more convenient method of reducing costs than mechanization. Miners, moreover, were generally opposed to measures that would reduce employment.

All in all, for reasons no more than suggested by the brief account above, throughout much of its history the bituminous coal industry has been unstable and disorderly to such an extent that it was commonly called a "sick industry." The best account of its difficulties is to be found in the 1925 report of the United States Coal Commission and the technical papers on which the report was based.[2] The troubles of the industry were naturally made more serious by the progressive intrusion of the petroleum industry into its markets. We have given the above account entirely in the past tense because, through a series of stages, the industry has been able gradually to eliminate the extreme disorderliness which had marked its history. In this process the federal government and the trade unions have played prominent parts. New problems, however, have arisen to plague the industry.

Federal Regulation and Its Aftermath

The first stage in the rehabilitation of the industry began with the imposition of federal regulation in 1933. This occurred under the National Recovery Administration, an experiment applied to many industries as a part of the New Deal policies of the Roosevelt Administration. Prior to

[2] U.S. Coal Commission, *Report of the United States Coal Commission* transmitted pursuant to the act approved September 22, 1922 (Public No. 347), Parts I-V, Senate Document 195, 68th Congress, 2nd session (Washington: Government Printing Office, 1925).

1933, a succession of Congressional bills had proposed various forms of federal regulation for the industry; but it took the extreme tribulations of the early depression years, and the changed political philosophy to which they gave rise, to induce positive action. Under the NRA, the Code of Fair Competition for the bituminous coal industry attempted to place the industry under a comprehensive system of minimum price regulation, mainly carried out by regional industry committees under federal supervision. The Code also gave the unions a new status in the industry—a status which in the end was to be the most important residue of the NRA experiment.

When in 1935 a decision of the Supreme Court struck down the whole NRA code structure, the coal code was one of the victims. Immediately, however, steps were taken in the Congress to supply a legislative substitute for the coal code. The Bituminous Coal Conservation Act of 1935 (the first of the so-called Guffey Acts) was struck down by the Supreme Court. The second Act, passed in 1937, included the former code provisions against unfair competitive practices, and provided for a system of minimum price fixing administered through district boards under the supervision of a national commission. After the complicated procedures for determining the price structure, the first official prices were introduced in 1940. But they served no purpose, since war demand carried coal prices above the minima. The Act, expiring in 1941, was renewed for two years and then allowed to die. Experience under the Act demonstrated the extreme difficulty of imposing price standards in highly competitive markets in the absence of effective control over production.

Following the bonanza situation created by high production and employment during the war, the postwar situation tended to degenerate into the state of industrial disorder characteristic of earlier times—a state somewhat modified by strong union organization inherited from the New Deal policy, but exacerbated by the rapidly increasing invasion of markets by oil and natural gas. A critical turning point came in the early 1950's—the so-called "turn-around"—when the United Mine Workers, under the leadership of John L. Lewis, seeing the handwriting on the wall, ceased to attempt to protect employment in the industry and started to co-operate with owners in making production more efficient. The resulting mechanization of the mines has led to a radical decrease in the labor force dependent on the industry. The average number of men employed is estimated to have declined from 442,000 to 134,000 between 1949 and 1965. At the same time, by reducing the unit cost of coal, it has enabled coal mining to become one of the highest paid employments in American industry. Average output per worker increased over the same period from 6.43 tons per day to 17.52 tons. Under the new union policy of trading employment for high wages, there has been a long period of years without major strikes; and the rising efficiency in production methods, apart from supporting high wages, has been the mainstay of the industry in preserving some of its markets from competitive fuels.

One consequence of declining employment was the emergence of a great social problem for Appalachia—to provide other employment in the region and halt the deterioration of community life. This problem still exists. Looking forward, however, the labor problem of the coal industry may be turning from one of excess manpower to one of shortage over the next decade or two. Difficulties in replacing the present labor force might limit the rising level of coal production.

A review of coal-mining labor, its age, skill requirements, and geographic structure, and the prospects for recruitment would be useful to permit a better judgment of future coal production costs.

Similarly, some of the difficulties due to the instability imparted to the industry by the traditional prevalence of small, inefficient firms are being replaced by questions arising out of mergers, both within the industry and with energy and non-energy firms outside the industry, that have occurred in recent years. If the trend is to "energy companies," which mine coal, lift oil and gas, and perhaps have a foot in the nuclear camp, then a new look at the conditions, and especially the cost, under which coal might be produced in the future is called for.

What are the implications for the future cost of coal of the increasing importance of very large coal companies? What is the impact upon the competitive power of coal of mergers in which coal companies become associated with producers of other energy sources?

Coal Reserves and Resources

As indicated earlier, the coal deposits of the United States are known to be vast, and remain one of the nation's most important potential energy resources. A recent estimate of the Department of the Interior of 220 billion tons minable at or below present costs works out to over 400 years' supply at present rates of production, and more than 100 times the present annual production of energy from all sources. Even if these figures are adjusted for future increases in energy demand, the estimated quantities would last far into the future. Moreover, it has been estimated that price increases of modest proportions would call forth large additional quantities. Quantities ultimately recoverable at unspecified price levels have been estimated at over 800 billion tons.

But much depends on how one phrases the question. In its 1962 report [3]

---

[3] U.S. Senate, Committee on Interior and Insular Affairs, *Report of the National Fuels and Energy Study Group on an Assessment of Available Information on Energy in the United States,* September 21, 1962, Senate Document No. 159, 87th Congress, 2nd session (Washington: Government Printing Office, 1962), p. 82.

the National Fuels and Energy Study Group of the Senate Committee on Interior and Insular Affairs reported on a special survey conducted for it by the National Coal Policy Conference. Coal operators were asked what they considered to be reserves at 1960 prices. Their collective replies, adjusted to represent the entire industry, yielded a figure of 20 billion tons, with an additional 15 billion coming in at a price increase of 25 cents per ton. Obviously, there is an enormous range in which estimates may fall, calling for the utmost clarity in concept and specifications.

There are considerations which, despite the impression of abundance, call for a new look at coal reserves. Some are discussed immediately below, others in subsequent sections.

1. Ownership and Commitment. The estimates of coal reserves need to be complemented by a more limited accounting of relatively short-run availability based on existing ownership and types of commitment. This would start with an estimate of reserves on a basis somewhat analogous to the concept of "proved reserves" in the case of crude oil, fanning out into larger amounts on the basis of relaxed economic and technological criteria. Within this accounting, certain types of long-run, large-scale commitments to given uses should be shown. The status of coking coal, captive and non-captive, could usefully be analyzed from this point of view. Is the steel industry protected from a future availability crisis? Similarly, large-scale contract commitments for electrical generation should be known and availability problems for future commitments analyzed.

2. Geography. Reliance on abundant coal reserves is usually couched in terms of large aggregates without much attention to location. Competition first with natural gas and now nuclear power for electrical generation, however, requires either that transportation cost of coal be kept low or that it be replaced by having the generation take place at the coal site (minemouth generation) and the power sent to consumers by wire. In either event, deposits within reasonable distance of consuming centers are preferred. In addition, these deposits must be large in order to permit economic utilization of costly transport systems that have to be provided or, even more so, to justify erection of minemouth plants. And, finally, coal that can be economically mined only with the simultaneous emergence of undesirable effects upon the environment may some day in the future lose its characteristic as a portion of the country's reserves. *With main consuming centers east or not far west of the Mississippi River, and thus in areas that have had their coal deposits heavily worked, a new look at coal reserves is needed.*

Several research questions come to mind. Where are the sites at which large blocks of coal could permit recovery of millions of tons per year

needed to serve large-scale power plants? At what prices could specified
amounts of coal become available at these sites? What transportation
costs would arise in their use for specific markets? What are the trade-
offs between coal and electricity transport from large coal deposits that
can be identified? To what extent will constraints on the degree to which
adverse environmental effects are allowed to occur shift the location of
coal mining and remove significant quantities from the reserve category?

3. Quality. While location and size need to be looked at in order to
get a more realistic grasp of the reality of coal reserves, *there is now an*
additional factor that needs consideration, and that is the sulfur content
of coal. Given the growing concern for air pollution, especially in urban
areas, sulfur removal from either the fuel or the stack gases is becoming
mandatory at various levels of concentration. Because removal is currently costly—and may remain costly for a long time—coal that has a low
sulfur content to begin with is much sought after. But hard information is
difficult to come by.

A careful look at coal reserves in terms of sulfur content, and its rela-
tion to potential markets in terms of distance and transportation costs,
merits high priority. Specific questions are framed in Chapter VIII.

Coal Costs and Prices

In earlier chapters on crude oil and natural gas, and earlier parts of the
present chapter, we have noted the degree to which coal has been supplanted as the principal source of energy for the American economy. It
has almost completely lost out as the fuel for space heating, railroading,
and water transportation, and has lost much of the market for industrial
fuels. Apart from metallurgical uses, almost the only market in which it
remains over wide regions the principal fuel is that of generating electrical energy. Even here, though, serious competition—from natural gas, to
a smaller extent from imported residual fuel oil on the East Coast, and,
yet to come, from the recent arrival of nuclear power above the competitive horizon—limits its potential.

These competitive threats have added a strong impetus to introducing
further measures to lower the delivered price of coal. Additional mechanization of the mines is one such measure, though in the case of surface
mining the resulting environmental damage and cost of restoration are introducing offsetting costs.

A complicating factor in assessing the future competitive outlook for
coal is the absence of useful information on the structure of coal costs
and prices. This may seem peculiar as the Bureau of Mines each year
publishes values of coal mined, by county, but these are averages of dif-

ferent qualities, mined by different methods, from mines differing in age, topography, equipment, etc. These value figures combine steam and metallurgical coal, so that it is impossible, for example, to learn minemouth prices of coal destined for utilities, and compare these meaningfully with cost of coal reported to the Federal Power Commission by the country's utilities as paid at the generating site. (Reporting to the FPC the price of such coal at the minemouth would be a long step ahead in gaining better knowledge.) Comparisons designed to show the role of transport cost are almost always limited to the national average minemouth price and the national average net transport revenue per ton.

Another facet that requires analysis is trends of and relationships between spot prices and those paid under long-term contracts. The increased role of long-term delivery contracts for very large plants makes such knowledge increasingly useful in any appraisal of competitive conditions.

Finally, as in the case of oil, we know next to nothing of the short- and long-run marginal costs of coal, neither for all coal, nor, as would be desirable, by type of operation and other characteristics. Such knowledge would help throw some light on the fact that coal mining statistics show a substantial portion of production originating in very large, newly opened mines. Are these the lowest-cost sources as production expands? Just how does the industry respond to rising demand? Current information is not available to answer such questions.

All in all, a determined effort is called for to collect and disseminate data on prices and costs, specific as to differences in quality, type, and age of mine, destination, ownership structure, terms of sale, and perhaps in other respects.

The Cost of Coal Transportation

Very important to coal's future competitive position are measures designed to lower the transport cost element that bulks so large in delivered price. Of these, moving coal by pipeline and designing railroad equipment and operations to cut unit freight cost are the two most prominent.

A coal pipeline has operated successfully but briefly in Ohio. Its operation, in fact, giving rise to plans for pipelines elsewhere, has often been thought to have been a potent factor in causing the railroads to modernize their coal-carrying technology and administration. But its very success was its doom, since it induced the appropriate rail lines to make rate offers to the utility in question that were sufficiently attractive to overcome the economies of the pipeline. A second pipeline, this one to carry coal from northeastern Arizona to southern Nevada, is soon to come into existence, however, and thus the coal pipeline will once again be more than just a technological potential, though hardly on the scale that seemed in the offing in the early 1960's.

Two factors brought to a halt the movement toward coal pipeline transportation that seemed to be gathering momentum in the early 1960's. One was the opposition of the railroads and, to a lesser degree, of mine workers, who feared that the transition to pipelines would lead to further concentration of production by making it difficult for the many small mines to gain access to the prospective pipelines. The second element, though not wholly independent, was opposition by several state legislatures to granting the necessary rights of way. Attempts in the Congress to deal with the matter on a national scale, initiated early in the Kennedy Administration, soon lost steam and were dropped altogether when rail transportation cost began to be reduced.

Coal pipelines have stood the test of technical and economic feasibility, but only to serve specific consumers. Their experience will not yield the information needed to judge a pipeline's most important merit, viz., economies of scale that have made them a preferred carrier both for crude oil and natural gas. Coal pipelines that, like railroads, function as common carriers, could achieve scale economies of a different magnitude, and studies of that economic characteristic could usefully supplement the research needs cited above with regard to location of large blocks of reserves.

Assuming that the issue will sooner or later again become crucial to the establishment of pipelines, a study devoted to the role of rights of way and possible institutional approaches would usefully supplement the topic outlined in the preceding paragraph.

A second cost-reducing innovation has been the unit train, which in its purest form is a shuttle train dedicated to service between a mine and a power plant (or other large consumer), able to load and unload at great speed and carrying severe penalties for failing to do so, and made up of cars especially engineered to accommodate coal (or, generally, any given commodity). Such trains have now been in operation on many lines for several years and since 1962 have reduced gross freight costs—i.e., not considering new costs due to charges for new equipment—by one dollar per ton and more.

The advent of the unit train has been possible only through the approval by the Interstate Commerce Commission of special rates, and there are some who see its emergence more as a product of regulatory than of technological innovation. To others, recent events suggest that perhaps giving the ICC authority to move on its own motion, rather than be merely responsive to proposals made by others, might be indicated.

As it has historically developed, the unit train has brought with it a new relationship between coal mine, railroad, and electric utility. For a variety of reasons, in some of the major instances the utility rather than the railroad has purchased and owns the necessary rolling stock. On rare occasions, the coal company has assumed that role. In any event, railroads have tended to limit their participation to one of operating the freight movement. This raises three classes of issues that need research:

1. *The first is an issue of facts, i.e., what are the true transportation rates, taking into account those portions now hidden in utility and, on occasion, coal mine accounts? What have been relative savings when all costs are considered? Thus, what is the true competitive situation vs. other means of carriage? And what is the prospect for further cost reductions in transportation and handling?*

2. *Does the assumption of large capital costs and obligations in the movement of fuel affect the character of electric utility operations? Specifically, does it affect the rate base and, if so, what are the implications?*

3. *Does the switch from buying transportation services to owning them reduce the degree to which utilities will in the future take advantage of changes in relative fuel prices, and buy in the cheapest market— especially as the economics of unit train transportation requires high-capacity utilization? Is the increased degree of being "locked-in," not only to a specific fuel but also to a specific location and means of transport, a factor that might delay introduction or diffusion of new technology?*

Public Policies Affecting Coal

Coal is the only energy source not now subject to direct forms of public regulation (other than in such areas as safety and environmental controls). Nevertheless, its competitive position has been much affected by regulatory practices in other industries. As mentioned, its position in various markets is partly defined by the structure of railroad rates regulated by the Interstate Commerce Commission. Federal Power Commission regulation of wholesale and field prices of natural gas helps to establish the terms of the price competition in many local markets. State regulation of the production of crude oil affects the terms of the price competition with oil, on the whole favorably to coal since output restriction lowers the quantity and raises the price of the competing fuel at least in the short run. The federal policy of restricting oil imports also favorably affects coal, though at the most crucial point of competition, the East Coast market for residual fuel, the restriction has been removed in all but the letter. Federal activity in nuclear research and development has been of great help in bringing nuclear energy within competitive range. Federally sponsored hydroelectric projects also have some effect on the market for coal.

A new governmental activity, concerned with the regulation of environmental pollution, affects all three fuels, but especially coal and oil. The effect upon the evaluation of coal reserves has already been mentioned; the wider ramifications are discussed in Chapter VIII. There remains to be mentioned here the pollution effects of coal mining itself. These arise principally through acid mine drainage and land despoliation, and have been the subject of regulation on a state level. An interstate compact that would set up standards applicable to the participating states has not as yet found sufficient adherents to be effective.

The issues associated with water and land pollution caused by coal mining have been well ventilated, and an extensive literature exists. Various states have been gathering experience with legislation aimed at minimizing land and water degradation. The time is ripe to search the records for answers to several questions. *How has such legislation affected the price of coal at the minemouth? How has it affected the competitive position of coal as between different locations in the same state and as between different states? Is there any notable decline in coal production in areas that have come newly within the scope of new regulations? Given the competitive outlook for coal, what tolerance limits, if any, can be established for preventative and restorative measures before costs become intolerable? How do the actual benefits from efforts at restoration compare with those expected and planned?*

Another direction from which public policy has begun directly to bear on the future of coal is through the expenditure of federal funds for research and development designed to find new uses for coal. The magnitude is exceedingly small when compared to governmental investment in nuclear R&D, but sizable in terms of R&D funds generated by the coal industry itself, which, in the past, has devoted its energies above all to the defense of its position as a supplier of energy to traditional users.

Among potential new uses of coal are liquefied and gasified coal, one to supplement oil, the other to supplement natural gas. After that comes the use of coal as feedstocks for petrochemical production—not new, but without a doubt capable of great expansion and eventually of substituting to a large extent for feedstocks now derived from crude oil and natural gas.

The first order for research in this area is to compare meaningfully with competing products, starting with the current state of the art, the costs of liquefied and gasified coal, including presence or absence of external costs (e.g., coal gasified at the mine eliminates anti-pollution costs at a later stage). How do assumptions regarding cost of raw materials (coal, gas), capital, incidence of taxation, and other factors affect the outcome? What is the effect of scale? What size coal reserves, in any location, would be required to satisfy appropriately scaled operations? What coal price range could various processes tolerate? Where are such reserves, taking into account the best-suited physical and chemical properties, and what are the locational implications? On what grounds should location of conversion plants be determined, and who should determine it?

Given the public utility aspects of gas transmission and distribution, could the price of gasified coal be left to the market? If not, would coal markets develop a dual pricing system?

V

Electricity

ALMOST EVERY CORNER of the electric power industry is subject to some form of government regulation or control. Government influence developed in a variety of ways in a changing historical environment and in response to different problems and philosophies. To understand the origins of the many forms of public intervention into the affairs of the electrical industry, it is necessary to review several stages of industrial and social history.

State Regulation

The necessarily monopolistic character of local electric utilities was recognized early in the history of the industry. Because of unsatisfactory early experience with private operators, around the turn of the century there was a considerable trend towards public ownership at the municipal level. But the pioneering efforts of Wisconsin and New York in 1907, in setting up relatively effective state regulatory commissions, turned this tide. State regulation of local rates and services has become the most settled part of the relation of public authority to the electrical industry.

State regulation in American law centers around the concept of a "public utility." Based on the English common law concept of activities "affected with a public interest," it has been developed by judicial decisions to determine what business activities may be publicly regulated because the public interest is not sufficiently protected by the operation of competitive forces. The attainment of public utility status by an industry entails the combined characteristics of (1) providing services of special importance which (2) are provided under circumstances that lead to monopoly or a highly wasteful or ineffective operation of competition. Electric utilities fall into this category because of the high ratio of fixed to total costs and the economy of having all users served by a single source. Competitive sources of service would be extremely wasteful of capital investment and would multiply the costs to be met by consumers.

As a regulated monopoly, the electrical utility normally has an exclusive service area. It is held to specified service standards and required to serve all customers. In practice, the heart of the regulatory process has been control over the rate of return. This has had two aspects: the general level of rates and the detailed rate structure. The general level entails the fixing of a rate base, or capital value of the assets, upon which the company is entitled to earn a specified rate of return; and the detailed rates, generally set to cover the assignable costs to each customer class plus an allocated portion of the joint costs, are designed to produce an amount of revenue sufficient to provide this rate of return after meeting all authorized costs. The legal rules applicable to determining the rate base and the rate of return have been determined by the courts, and ultimately the Supreme Court, through a long historical record of litigation. In a general way, the rate base is normally attached to capital investment with various adjustments; and the idea of a "fair return" is related to a competitive norm—an amount necessary to induce the desired investment. But the problems are complex in nature; and the body of judicial reasoning surrounding them is prodigious. The structure of rates involves setting a number of different rates for different classes of customers. To a degree these attempt to assign differential costs to various groups; but they also take advantage of different group demand elasticities for the purpose of stimulating sales and improving load factors. Regulation is charged with seeing that such rate structures are not "unreasonably" discriminatory.

The comprehensiveness and effectiveness of state regulation have varied greatly from state to state. The holding company phase of corporate control in the 1920's revealed the inadequacy of the powers of most commissions; and in the early 1930's many state commissions were reorganized and given additional powers. Given the complex of municipal, business, and consumer interests involved, the political byplay leads to very different results in different states.

Increasingly, the traditional concentration of regulation on rates of return, rather than on efficiency of electric industry operation, has been subject to criticism. This is one of several basic questions regarding the purposes, efficacy, and criteria for regulation on which research is suggested below (pages 92-93).

The Involvement of the Federal Government

Beginning with the 1930's, the federal government came to play an increasingly important part in the ownership, operation, and regulation of electric power facilities. Prior to that time its involvement in electric energy had been mainly an indirect result of its concern with water resources. Hydroelectricity was a by-product of the storage dams built by the Bureau of Reclamation to bring irrigation water to the arid West; therefore, provision had to be made for the sale of electricity and for the use of revenues from its sale. Similarly, the Federal Power Commission, created in 1920, was given the authority to license non-federal hydroelectric projects on streams subject to federal jurisdiction, largely out of a concern with the broader objectives of river development.

Federal electric power policies, as such, are essentially a product of the 1930's. This decade, in particular the years of Franklin D. Roosevelt's New Deal, saw the adoption of several major laws seriously affecting the operations of the electric power industry. These laws resulted from the conjuncture of such broad forces as the Great Depression, with the consequent drive for economic reform, and specific forces growing out of the changing economic and technological circumstances of the electric power industry.

Just as, in the early 1900's, the demand for state regulation came about to overcome the deficiencies of regulation at the local level, in the late 1920's and early 1930's the need was felt for federal regulation to supplement regulation at the state level in dealing with those aspects of electric industry operations which proved to be beyond state regulatory control. The electric power industry had in many ways outgrown state boundaries. In no way was this more evident than in the growth of holding companies, especially during the 1920's.

THE HOLDING COMPANY PHASE AND ITS AFTERMATH

During the 1920's, a large proportion of the local companies fell under the control of a small number of holding company groups. This trend had started in a small way around the turn of the century. The experience of the General Electric Company is a case in point. As manufacturers of electrical generating equipment, the company promoted its sales by its willingness to take stock in operating companies in part payment for new equipment. A special subsidiary company was set up to administer these investments, the Electric Bond and Share Company. Having this interest, the company found that by providing technical and management advisory service to improve the operating efficiency of the local operating companies, it was possible to make them much more valuable. Though it seldom held a majority of the voting stock, the Electric Bond and Share Company was in a position to influence company managements. The policy having proved a profitable one, it was extended to a deliberate policy of acquiring stock of other companies. Apart from gains to be made through improved operating efficiency, income could be gained from the sale of advisory services; and the company connections were useful as an aid to the General Electric Company in selling equipment. Moreover, in the constant struggle and litigations over the valuation of producing properties and the permissible rates of return, legal and other technical assistance could be rendered to companies in their dealings with state regulatory commissions.

A similar pathway was followed by other interests among whom, for example, was Stone and Webster, a management consultant firm which began by taking stock in companies in part payment of fees, but later extended the range of its operation. Eventually the possibilities attracted members of the financial community, such as investment bankers, underwriters and promoters, who began floating new companies. In the bonanza days of the 1920's these activities reached colossal proportions. By 1929, the seven largest holding company groups accounted for some 60 per cent of electrical generation, and there were numerous smaller groups. The opportunities for profit were of two sorts: (1) an improvement in the earnings of the operating companies; and (2) the sale of management, legal, financial, and technical services.

On the first of these points, the times were propitious. In the booming business situation of the 1920's, earnings were in any case on the upswing. Also, possibilities for improved operating efficiency general existed. On a lower ethical plane, because of the inefficiency prevailing in most state regulatory agencies, augmented by judicial rulings, effective action could be taken to keep rates high.

On the second point, the services for which the holding company charged might, or might not, be valuable. In any case, they represented costs to the operating company which were beyond the control of state public utilities. The holding companies organized out-of-state subsidiaries to perform the services. In the case of some holding company groups, the service charges were excessive to a scandalous degree.

In many cases the corporate and financial structures of holding company groups attained a preposterous measure of complexity. By the process known as "pyramiding" one holding company was piled on top of another, sometimes to four or five levels, and subsidiaries fanned out in all directions. By financing at each level partly by fixed income securities, the owners of equity stocks received pyramided profits from any increase of income at the base; and those controlling the topmost company could control a vast "empire" of underlying operating companies. At the base of the structure, the operating companies were heavily financed by fixed-interest bonds, so that any increase of income accrued to stockholders and flowed upward to greatly magnify the earnings of those with equity shares in the superstructure of holding companies.

This happy system of getting rich quick held together so long as utility earnings were on the rise. But it fell apart rapidly with the stock market crash of 1929 and the deepening depression in the early 1930's. Lower earnings of operating companies slowed the flow of funds to meet the debt charges and dividends at the higher holding company levels, sending many of them into bankruptcy and spreading disaster to large groups of investors. Some groups, more conservatively managed and less hysterically financed, suffered less serious losses than others.

To pick up the pieces of this financial shambles, the Congress passed the Holding Company Act of 1935. The Securities and Exchange Commission, which had been set up as a New Deal agency to supervise the financial markets, was empowered not only to regulate the financial operations of registered holding companies, including the acquisition and disposal of properties and assets, but it was also given the power to break up existing holding companies, except for those confined to a single interconnected system serving a contiguous territory. This onerous task extended over a period of many years.

During all this history the electrical utility operating companies had gone about their business of generating and marketing electricity; the disorder had taken place above their heads in the world of financial manipulation. Many of them may have come out of the experience in a better state of technical efficiency than they entered it, since it had been to the advantage of the holding companies to improve the underlying companies. Naturally, many companies had difficulties in making financial adjustments. Some had been over-financed and had excessive capital charges. Some had been drained of their cash assets. Many top managements had to be replaced. But, generally speaking, the companies were in a position to continue satisfactorily the performance of their productive functions.

The electric holding company systems that today operate under the provision of the Holding Company Act account for about 20 per cent of the electric energy sales of privately owned utilities. Since the 1930's many other privately owned utilities are said to have refrained from entering into jointly owned generating and transmission operations that could be justified on economic and technological grounds, for fear that such arrangements would make them subject to the jurisdiction of the Securities and Exchange Commission under the Holding Company Act.

The holding companies undoubtedly accomplished some good in achieving a necessary degree of co-ordination in an industry whose operations were being driven across state boundaries by technological and economic circumstances. But this was cancelled out in the eyes of the public by the financial collapse of many utility holding companies. The utility holding company became identified with ruthless financial manipulations. *The legacy of this period in the history of the electricity industry persists today as the urgent task is faced of adapting the structure of the industry to technological advances which require operations on a vast scale if the utmost efficiency in generation and transmission is to be achieved. This problem is addressed in the research recommended on pages 97 ff.*

The 1930's also saw federal supplementation of state authority in the regulation of utility rates. Thus, legislation passed in 1935 included among its provisions authorization for the Federal Power Commission to regulate the rates charged for electricity sold at wholesale in interstate commerce. This legislation was designed to close a gap in rate regulation that emerged as some electricity began to be sold across state lines, and hence outside the control of state regulatory authorities. This grant of power to the Federal Power Commission remained largely dormant until the 1960's. This aspect of federal involvement, which is of great potential importance, is dealt with below on pages 91 ff.

FEDERAL ENTERPRISE

The deep and direct involvement of the federal government in the field of electrical power came as an element in President Roosevelt's New Deal program in 1933. Prior to that time, as noted earlier, a number of small hydroelectric projects had been developed by the Bureau of Reclamation of the Department of the Interior in connection with irrigation projects. The power was sold at the damsite to local distributing companies, largely on their terms, and the transactions raised no controversial questions in principle.

The first big test of principle came in the 1920's in connection with the use of the waters of the Colorado River. The City of Los Angeles and the Southern California Edison Company had been competitors for damsites on the river to fortify their positions in the rapidly expanding electricity markets of Southern California. To commit the waters for the use of dams solely to produce electricity for Southern California would, however, have contravened the responsibilities of the federal government. The waters of the Colorado River were one of the principal potential economic assets of the seven states in the river basin. Some principle had to be evolved for dividing the beneficial use of the water among them, a problem worked out in the terms of the Colorado River Compact, signed by the states and approved by the Congress. In addition, international questions were involved, since the lower river ran through Mexico which possessed rights in the waters. The federal government therefore retained control of the river development.

As a first step, it built the Boulder Canyon (Hoover) Dam which, together with other installations further down the river, was capable of serving all the purposes. The government itself did not engage in electrical generation. It installed the generating equipment and leased its use on long-term contracts, mainly to the City of Los Angeles but in minor part to a private utility company. The revenues under the contracts were sufficient, not only to pay off the capital costs assignable to electricity, but also to pay off a large part of the cost of the total installation.

Federal enterprise in river basin development on a large scale, including electrical installations, began in 1933. The Boulder Dam project had stood alone, arising ad hoc out of particular circumstances, and not reflecting any general policy or program. The later projects were elements in a broad range of policies of the Roosevelt Administration comprehended under the name of the New Deal. The New Deal arose out of the deep economic depression and embodied a new philosophy of the relations between government and private business. In the case of electricity, the times were propitious for public action because of the abuses that had arisen under the holding company regime and the financial debacle that had overtaken the industry.

The first and most ambitious of the federal river basin projects was the Tennessee Valley Authority. It was the culmination of a long political struggle over private vs. public power on the Tennessee River. During World War I the government had built the Wilson Dam at Muscle Shoals for the manufacture of nitrogen for war purposes. In the years after the war, various proposals were made for selling or leasing it to private interests for purposes of electrical generation and the manufacture of agricultural fertilizers. While the political opposition successfully resisted these proposals, it did not succeed in establishing a constructive federal program of development. Standing as a symbol of failure to develop a great reserve, Muscle Shoals became the rallying point for forces urging federal development of river basin resources.

For this purpose, the setting up of the TVA provided the opportunity for a program embodying a wide range of New Deal objectives. In its broadest aspect, it was a program for the regional development of an economically backward region, based on utilizing the full potential of the waters of the Tennessee River basin for a variety of purposes. In its narrowest aspect, it was a means to provide emergency employment and to stimulate industrial activity. Concerning its specifically electrical aspects, they were in part designed to promote the first of the objectives mentioned above, by inducing the entry of light industry into the region. On another front, they were designed to demonstrate a theory about electrical rate structures, namely, that the demand for electricity was highly price elastic and that the use of electricity could be profitably expanded on the basis of rates much lower than those typically charged by private companies.

To build up the market for its electricity, the TVA offered preferential service and low rates to municipally owned dstributive systems, and federal assistance was given to municipalities either to purchase existing systems or to build new ones. It also gave preference to rural co-operatives and federal assistance was provided for setting up such systems. The TVA area provided the experience upon which the whole later development of rural electric co-operatives was predicated. In the outcome, TVA supplanted private power supply sources and became almost the sole source of supply for a wide geographical area in six states. To meet the rapidly mounting demand, it has since built steam-generating stations as well as hydroelectric installations. Today the majority of its production is from steam-powered plants. By progressive expansion it has become the country's largest electric system.

The arguments advanced in favor of the TVA could equally be advanced for the development of other river basins, especially in the Far West where great streams ran unused through arid and industrially undeveloped regions. This was particularly the case with the Columbia River basin where great volumes of water flowed through gorges well suited to high dams, to support irrigation, electrical generation, and related industrial development. Strong arguments also could be advanced against such projects. One was that the market did not exist for a great increase in the amount of electricity, so that revenues from the sale of power would not cover the costs attributable to power alone, to say nothing of helping to amortize other costs. Another was that, in an agricultural situation in which farmers elsewhere were paid to remove land from cultivation, it made no sense to subsidize the development of great areas of arid land.

But the projects were supported, not only by strong regional political pressures, but by the expansionist philosophy of the New Deal and by the belief that a tremendous growth in the electricity market could be achieved through low rates. In any event, the projects were started and have been continuously expanded and increased in number.

A significant aspect of the direct involvement of federal enterprise in hydroelectric power centers on the promotional low-rate policy which was introduced by TVA and followed in other projects. This is regarded by many as a turning point in the history of the electrical industry. When in 1933 the federal government entered the hydroelectric business in a big way, it also adopted an active marketing policy. It was felt that private utility rates were unnecessarily high, that sales could be greatly stimulated by lower rates, and that these would be economically justified by higher consumption in view of the influence of diminishing costs in this capital-intensive industry. Consistent with these views, a commensurately low level of rates was imposed upon purchasers distributing at retail. The use of low federal power rates as a yardstick for comparison with private utility rates was also expected to bring pressure on private companies to lower their rates.

The "yardstick" idea could be attacked on the ground that in multipurpose federal projects the costs assigned to electrical generation were arbitrary and also tax-free, and thus not comparable with the costs of private companies. This is in principle a valid argument. But it did not prevent TVA, the Bonneville Power Administration, and other projects from setting out to demonstrate that a vast market existed for low-priced electric power. It is argued by many that the point has been proven and that benefits have accrued to users of private power in two ways: through local pressures for lower rates, and through a recognition by companies that low rates could be profitable. We suggest this as an important subject matter area in which academic research should be performed. The first objective of such research should be to test the validity of the basic argument (which would include attention to the short- and long-run price elasticity of the demand for electric energy, as well as attention to the long-run cost curve of electricity supply). If valid, it may be of critical importance for the future of regulatory practice and indicate the need for additional research along the following lines:

1. *Will aggressive federal enterprise, or other forms of publicly owned enterprise, continue to be an essential supplement to the activities of regulatory agencies in the future? The dynamic character of the industry in its technology, supply, and demand characteristics may or may not indicate the continuing need of effective competition from publicly owned enterprise as an adjunct to regulation.*

2. *With the diminishing relative opportunities for hydroelectric power as an element in electric power supply in the future, will there be sufficient opportunity for federal enterprise to perform the role it is said to have played in the past? If not, will the full burden fall upon regulation as such, including the newly revitalized regulatory role of the Federal Power Commission, discussed below? What new criteria and standards for regulatory practice would be indicated?*

RURAL ELECTRIC CO-OPERATIVES

Next to engaging in direct electrical enterprise, the most important influence of the federal government has been exercised through sponsoring rural co-operatives. In 1931 only about 10 per cent of farms were served by electricity; today virtually the whole rural population is served, about half by co-operatives, the other half by private or non-federal public agencies whose entrance into the field was largely stimulated by federal programs.

Active sponsorship first began in the TVA area where co-operatives were among the preference customers specified by the TVA Act of 1933; and TVA was also authorized to construct transmission lines to rural areas and villages to encourage electrification. The Rural Electrification Administration was subsequently established in 1935 to promote and administer the program in which federal loan funds were made available at low rates of interest, and in which the co-operative was used as the main unit of local organization. Private companies were offered federal loan funds, but given the temper of the times the response was so slight that the co-operative became the chosen instrument. In many states special laws had to be passed, under the spur of REA, to provide a legal framework for co-operative ventures.

Under the loan contract between REA and the borrowing co-operatives, REA has the authority to supervise major aspects of operating policy, including approval of contracts to purchase electricity. Authority of REA to oversee retail rates is included in the contracts, but is subordinate to state regulation in some states or to resale rates specified by federal sources of power. Rates are expected to cover debt service (interest and amortization), taxes, operating costs, and reserves. Uniform accounting based on FPC standards is required. Freedom from income taxes and low-interest loans (at present 2 per cent), and their non-profit character, have enabled most co-operatives to keep rates low. Generally speaking, they have had a good financial record. Co-operatives are almost wholly engaged in retail distribution; out of nearly 1,000 in 1962, only 76 were engaged in generation and transmission.

The operations of co-operatives present a number of controversial features on which study is needed. Usually they do not have defined service areas, so that they may run into competitive overlapping with other distributors, private or municipal. Objections are also raised to the expansion into generation when pools of co-operatives organize special entities for generation and transmission (so-called "G and T's"), a growing phenomenon. Most G and T units are small by modern standards. However, they serve the purpose for which they were intended, i.e., to meet the power needs of the people who qualify for loan assistance under the Rural Electrification Act. Although they supply only one-fifth of the total co-operative power supply, they serve as a competitive threat to private companies and are valued by the co-operatives as a bargaining tool. How to fit them into an expanding network of co-ordinated systems also presents a problem. Opposition has been voiced to the continuance of low-interest loans to the co-operatives after electric service has reached practically all rural communities. The extent of the need for the present REA loan authority merits study in the light of possible availability of funds—or electricity—from other sources for meeting the basic objective

*of the program of furnishing dependable and efficient electric service to
rural people, now and in the future.*

FEDERAL REGULATORY ACTIVITY:
THE FEDERAL POWER COMMISSION

Apart from the projects that it directly operates, the federal government is involved with the electrical industry in a regulatory capacity in
three different ways: (1) It licenses the use of hydropower sites on
"navigable" rivers under federal jurisdiction. (2) It regulates the wholesale rates of power sold for resale in interstate commerce. (3) It is responsible for encouraging the interconnection and co-ordination of power
systems; this is not a regulatory function as such, but it is a responsibility
that is growing in importance. These functions are performed by the
Federal Power Commission. The first involves it in relations with federal
agencies directly engaged in hydropower development. The second involves it in overlapping and somewhat ambiguous relations with state regulatory agencies. The third is now in the process of evolution. (See the
final section of this chapter, on "Future Industrial Structure.")

1. Licensing of Hydropower Sites. The federal government first became involved with the electric power industry through issuing licenses
to private interests to develop hydropower projects on rivers subject to
federal jurisdiction. Licenses were at first issued by direct act of Congress. Early in the present century the terms upon which licenses should
be issued were the subject of acute political controversy. Private interests
sought rights with minimum restrictions. The "conservationists" wanted
strict regulation of site development at least, and many were in favor of
public development. They stressed the importance of multipurpose river
basin development to utilize the full economic potential as well as to
serve aesthetic and recreational ends.

The Federal Water Power Act of 1920, which set up the Federal Power
Commission to administer a general licensing policy, showed real concern
for the conservationist point of view. The FPC was authorized to issue
fifty-year licenses which provided for regulation of rates, service areas,
and security issues, unless they were subject to effective state regulation.
At first ineffectively organized under a commission made up of three
Cabinet members, for ten years it did little to plan for hydro development or to perform its regulatory functions. Reorganized in 1930, it took
its present form of five full-time commissioners with a permanent staff.

There were 279 licenses in effect as of January 1, 1964. Under the fifty-
year limit, a number of them are nearing expiration. At the end of the
period, the sites may be relicensed to the same or a new operator, or may
be recaptured, with compensation, for government operation.

No general policy has been formulated for dealing with expired licenses. This is an imminent as well as a long-term issue, and it requires, at the least, clear-headed review and study. The general approach to the correlation of the energy and non-energy aspects of water use has changed greatly since the original licenses were issued. The value of recreational and aesthetic aspects of water use have undergone profound change. The modification of licensed hydro sites well before the expiration of licenses in order to develop recreational potential is worthy of serious consideration. And even within the limited field of electric power production, the potential of hydro as a source of peaking capacity is becoming more valuable, particularly in connection with the large interconnected systems that are becoming typical of the organization of electric energy production. A series of case studies to throw light on the effects of these new factors may provide the best means of dealing with this question from a research standpoint.

The 1935 Act also gave the FPC broad authorization, never strenuously activated, for making studies upon which to base comprehensive river basin development, and to tailor particular licenses to the standards set up for each basin. However, its planning responsibilities along this line present a somewhat confusing overlap with those of other federal agencies, such as the Corps of Engineers, the Bureau of Reclamation, and the Geological Survey.

Many sites and projects are studied many times over by the same and different agencies. State and local agencies and private companies also have a stake in the planning activities. The Congress is under pressure from many quarters with respect to authorizing projects. The process of moving from planning to reality contains a complicating quirk: A federal project must be authorized by the Congress. The FPC, on the other hand, has authority to license power sites to private or non-federal public agencies. Therein lies the possibility of conflict of judgment among agencies as to licensing or reserving sites for federal development.

While several agencies may be involved, every federal project must go through the Congressional mill for approval and appropriation of funds, both for construction and for annual operations, since the revenues of the marketing agencies go into the federal Treasury. The Congress is thus able to exercise very close supervision over management practices as well as original authorizations. "Politics" is a pervasive feature of the considerations in most projects. This means that Executive and Congressional action is influenced by regional and party interests and by the representations of the private utilities. There is always a project waiting list and a competition for funds. The public vs. private power controversy has lost much of its ideological heat; but conflicting interests in particular projects can still generate political heat.

The problems created by the peculiar fragmentation and overlapping of study and advance planning for the utilization of water resources extends well beyond matters of particular concern from an energy standpoint. Furthermore, the problem seems to have been widely studied by political scientists, economists, geographers, and others. Consequently, no proposals for research on this problem area are here offered.

2. *Regulation of Interstate Operations.* The powers of the FPC were expanded under the Federal Power Act of 1935 to give it certain responsibilities for regulating companies engaged in interstate transmission of electricity, including the wholesale rates for power moved in interstate commerce. The individual states are unable to exercise regulation over the operations of companies transcending state boundaries as to rates, transmission facilities, or financing. After the Supreme Court decision in the Attleboro case of 1927 [1] demonstrated that the limitation on state regulation might become extensive, the Federal Power Act (synonomously the Public Utility Act of 1935) was passed to close the "regulatory gap." It was designed not to supplant but to supplement the regulatory powers of the states. The FPC has authority to fix "just and reasonable" wholesale rates, set standards of service, and prevent discrimination between localities or classes of service for electricity sold in interstate commerce.

The principle of supplementing state regulation does not, however, remove the fuzziness between state and federal jurisdictions. Comingled electricity (like comingled gas) raises knotty problems of overlapping authority. The FPC may investigate the cost of the property of every public utility with interstate connections, the depreciation practices, and other facts bearing on rate-making. It has the power to regulate accounting practices, some security issues, and some mergers and interlocking directorates. Since the same companies engage in intrastate and interstate operations, mostly intrastate, and state commissions have similar powers over intrastate operations, the state and federal regulatory powers may run into conflict, except as the respective jurisdictions are clearly defined by the federal courts.

The scope of FPC jurisdiction has been liberally interpreted in the courts. In 1942, a lower federal court upheld the FPC jurisdiction over the accounting practices of a company which operated locally but was interconnected with other companies selling beyond the state. The Supreme Court in 1943 further enlarged the jurisdiction, when it gave FPC authority over a purely local company which occasionally supplied power to another intrastate utility which in turn interchanged electricity with a utility company in a neighboring state.

---

[1] *Rhode Island Public Utilities Commission* v. *Attleboro Steam and Electric Co.,* 273 U.S. 83 (1927).

Though possessing wholesale rate-making authority, the FPC made little use of it before 1961. Interstate wholesale sales were a very small fraction of the business, and the FPC was much occupied with other matters, especially the regulation of natural gas rates.

In 1961 an invigorated FPC began to exercise this long dormant power. And in 1963, in the "City of Colton case," which had been making its way through the courts for many years, having been initiated by the city of Colton during the 1950's, the Supreme Court ruled that the FPC had jurisdiction over rates in sales made in intrastate commerce if a portion of the electricity sold originated outside the state—even though the sale was subject to the jurisdiction of the California Public Utilities Commission.[2] This and later actions of the FPC open up a wide vista of possible federal encroachment upon what had been regarded as the preserves of state regulatory authority.

The FPC could be placed at the very heart of the regulatory process in this country, particularly if the industry develops along the lines that appear to be indicated by changes in technology (the subject of the final section of this chapter). The regulation of rates, which in the past has been mainly the concern of state commissions, will increasingly involve a mixture of state and federal authority. The basic issues in regulatory economics on which research is needed will therefore increasingly be of direct concern to the federal government, as well as to the state regulatory commissions.

The research needs in regulatory economics are perhaps best stated in terms of the critical questions that have been raised concerning current regulatory practices.

1. *It has been argued by some economists that the rate-base system of regulation, in which the company is entitled to earn a specified rate of return on the capital value of its assets, leads to overinvestment. This is alleged to occur in a variety of ways, including the purchase of redundant equipment, the payment of high prices for equipment, etc. The hypothesis remains unproved and requires study.*

2. *A related argument is that the rate-base system of regulation provides no incentive to efficiency, and that study needs to be given to the development of regulatory practices that would encourage efficiency by permitting the more efficient operators to reap some of the benefits of their greater efficiency.*

3. *Another economic argument is that not enough attention has been paid to the structure of rates for different categories of consumption,*

[2] *Federal Power Commission* v. *Southern California Edison Co. et al.,* 376 U.S. 205 (1964).

and their effects upon the demand for electricity and ultimately upon capital requirements. In other words, the structure of rates may not be properly related to the true costs imposed upon the electric utility by the resulting demands. In this sense investment optimization and economically efficient prices are related problems; these problems, it should be noted, become more pressing as electricity is increasingly employed in newer uses such as space heating, air conditioning, and possibly private transportation.

4. *Another fundamental question pertains to the effectiveness of regulation, per se, in view of the growing competition among energy sources in so many of the applications in which electricity is employed. In addition, to what extent does the presence of some publicly owned systems serve as "workable" competition to the private utilities and vice versa?*

5. *What is the economic effect of the combination electric-gas utility? Does the combination result in lower aggregate costs? Or is its major effect rather to make the utility less aggressive in cutting costs and promoting growth than companies devoted to a single form?*

Directions for the Future:
Technology, Policy, and Industrial Structure

Technology, which from the beginning has been a dynamic factor in the development of the electricity industry, has now begun to assume a new importance in shaping the industry's future. Advances in techniques are increasingly extending the distances over which electric energy can be economically transmitted. Technological progress is also widening the cost advantages of extremely large-scale steam-electric generating units.

Transmission, although a moderate component in the total cost of delivered electricity, is of strategic importance in current and prospective developments. It is through the interconnections among electric power systems made possible by extra-high voltage transmission that economies resulting from large-scale generating units can be realized, as well as investment savings through sharing capacity to serve peak demands (which frequently occur at different times in different places) and to provide for reserve requirements.

Economies of scale point to bulk power supply networks, served by massive generating units, as the efficiency ideal for the future. Perhaps the major problem ahead is how to achieve the efficiency ideal in the face of the existing pattern of policies and the industrial structure supported by these policies.

THE PRESENT PATTERN OF POLICIES

As a result of the historical developments summarily described in the preceding pages, the present-day U.S. electric power industry is touched by public authority mainly in the following ways:

1. Regulation of the retail rates and services of private companies serving local communities is carried out by the states, usually through the agency of a publc utility commission, with some exercise of local authority in a few states.

2. There are numerous local systems owned and operated by municipalities which may, or may not, be subject to regulation by the state commissions according to the law of the various states.

3. The Federal Power Commission regulates the wholesale rates of electricity moving in interstate commerce, and exercises (mainly) advisory responsibilities with respect to the nationwide interconnection of facilities.

4. The FPC is also the custodian of hydropower sites, except those federally developed, licensing their use by private or by state or local public agencies, and exercising supervisory authority over their development.

5. The federal government operates numerous hydropower projects, usually as adjuncts to multipurpose projects of river basin development for irrigation, flood control, and regional economic development. In one large region, the Tennessee Valley Authority has become the sole source of supply for the region through a combination of hydro and steam plants.

6. In areas where the federal projects generate electricity, it is sold at wholesale to publicly owned distributing systems, co-operatives, private systems, and industrial users. Publicly owned systems and co-operatives are designated as preference customers.

7. Through the Rural Electrification Administration, the federal government provides financial assistance to co-operatives for extending electrical services to rural areas.

8. Some states, or subdivisions of states, operate publicly owned hydroelectric projects under federal license; and one state, Nebraska, contains a system of generation and distribution that is entirely publicly or co-operatively owned.

9. Through the Securities and Exchange Commission, administering the Holding Company Act of 1935, the federal government has effected a radical revision of the corporate structures of the private power systems.

10. The federal government has conducted a program of research and development, as described in Chapter VI, through which nuclear energy has been brought into competitive range as a source of energy for generating electricity.

While not exhaustive, this list covers the principal ways in which the electrical industry is controlled or affected by public authority.

PRESENT INDUSTRIAL STRUCTURE

The present pattern of industry to which the foregoing policies apply, and of which it is, to a certain extent, the product, is summarized in Table 12.

From this table, certain facts stand out sharply:

• Though the 480 investor-owned companies represent only 13 per cent of the electrical enterprises, they serve 79 per cent of the customers.

• The federal government operates 13 per cent of the generating capacity, but not in service of retail customers.

• Non-federal publicly owned systems (mainly municipal) make up a majority of the total, but they serve less than 14 per cent of the customers.

• Nearly two-thirds of the systems are engaged only in retail distribution, purchasing their power from the others.

• Over three-fourths of the generating capacity is in private hands, the rest in public or (minutely) co-operative hands.

Behind these basic facts lies a complex scheme of relationships, philosophies, and controversies to which we have already given attention.

FUTURE INDUSTRIAL STRUCTURE

The Federal Power Act of 1935 instructs the FPC to draw up regional plans for interconnection and co-ordination of systems. The *National Power Survey*, issued in 1964, contained the first such effort on a national scale. In this Survey, the FPC estimated that savings of as much as $11 billion a year are possible by 1980 as a result of technological progress, market growth, and full use of the potentialities of co-ordination. But the Commission also pointed out that, although the industry's pluralistic structure has provided a powerful competitive stimulus to cost reduction, "the large number of separate systems [there are 3,600 separate electric power enterprises], coupled with rivalries and controversies between segments of the industry, has frequently resulted in economically meaning-

TABLE 12. Number of Systems, Generating Capacity, and Customers Served by U.S. Electric Power Industry,[a] by Ownership Segment, 1962

Ownership	Number of systems			Generating capacity, per cent of total	Retail customers served	
	Total	Engaged in generating and transmission	Engaged in distribution only		Number	Per cent
Investor-owned [b]	480	318	162	76%	47,500,000	79.0%
Public (non-federal)	2,124	864	1,260	10	8,118,000	13.5
Co-operatives	969	76	893 [c]	1	5,095,000	7.5
Federal	44	42	2	13	–	–
Total	3,617	1,300	2,317	100%	60,713,000	100.0%

[a] Excludes Alaska and Hawaii.
[b] Includes 34 industrial concerns that supply energy to other customers.
[c] Many of the distribution co-operatives are also members of generating and transmission co-operatives (the so-called "G & T's") and hence participate indirectly in the generation and transmission function.

Source: Federal Power Commission, National Power Survey (1964), Vol. I, p. 17, Table 3.

less boundaries for utility system planning and operation which undoubtedly cost the power consumers of the country millions of dollars every year in wasted opportunities for cost reductions." [3]

The great challenge for the future is to find ways in which the different parts of the electric power industry, competitive though they may be, can "peacefully co-exist," all sharing in the benefits offered by advancing technology and passing on to the consumer the resulting cost savings. The major research needs, those to which the most urgent attention must be given, are in this broad area.

Basically there are three modes of adjustment by which existing systems can accommodate to increasing scale requirements: merger, co-ordination, and wholesaling. These are not alternatives, but each is part of an over-all pattern that is now in the making.

Mergers were the favored mode of rationalization before the 1930's, but their acceptability has been impaired by the excesses of the holding company period. In the 1960's there has been a resurgence of mergers on a modest but significant scale. *Research is needed on the circumstances and degree to which mergers represent a more efficient solution than co-ordination. To what extent, if any, are greater economies involved? To what extent can individual initiatives be relied on to provide a merger pattern consistent with the public interest?*

Co-ordination of systems has grown strikingly in the past fifteen years, with most of the major investor-owned systems and some of the public systems now joined in regional groups of varying "tightness" and effectiveness. *Systematic research is needed of the degree to which this process has generally approached the efficient ideal. Information needs to be collected and evaluated on the current state of power pooling compared with what could be achieved. A related question needing study is the appropriate role of regulatory agencies with respect to power pools: Should regulators get involved in an evaluation of power pools? Should regulators be able to require that systems join pools?*

Wholesale purchases are the means by which small systems, which for one reason or another are outside of power pools, can reconcile their independence with the requirements of efficiency. Success in this case will depend upon the ability of the small systems—public and co-operative—to buy wholesale power on equivalent terms to the cost of bulk power to the large vertically integrated systems. As "preference power" from federal hydro projects becomes scarcer, private wholesale sources will become increasingly important, as will effective regulation over wholesale rates. *Research is needed on the proper basis for the wholesale pricing of*

[3] Federal Power Commission, *National Power Survey* (Washington: Government Printing Office, 1964), Vol. I, p. 5.

electricity, along with a thoroughgoing study of the prospects and problems involved in wholesaling by the private companies to the public and co-operative distribution systems.

The three foregoing modes of accommodation require research specific to the problems of each. *But beyond this there is need for a fundamental study of the future structure of the electric power industry, which comes to grips with such factors as the outlook for the growth and location of electric power from federal installations, the availability of preference power, and the likely growth of the various ownership segments of which the industry is composed. Such a general study is needed to place mergers, co-ordination, and wholesaling into proper perspective. Such a study should also place in perspective the degree to which the requirements of an efficient pattern of growth will raise problems of an antitrust nature. It should also reveal the extent to which existing differences among the ownership segments with respect to taxation and financing (and resulting fixed charges) can prove a hindrance to effective co-ordination.*

Perspective is needed also on the reliability of power supply, a problem which was brought to public attention so forcefully by the Northeast power failure of November 1965. Designing and operating strongly integrated electric power systems so that they will be sufficiently reliable is not a purely technical problem but also rests on a series of economic decisions.

The social and economic losses sustained from an interruption in power supply far surpass the loss of revenues to the electrical utility. Study is needed as to the value of differing degrees of reliable service to various classes of consumers. Not only could information on this subject serve as a guide in the establishment of tariff rates for the sale of electric power, but, more importantly, it could provide a basis for determining the extra costs of power supply worth incurring for the purpose of assuring system-wide reliability.

Finally, with the more distant future in mind, it would be well to look beyond the present-day structure of the industry and ask the question whether a fundamentally different organization of the electric power industry might best serve the interests of efficiency. Do prospective changes in technology argue for a separation of functions in which generation and transmission for the entire nation are handled by a limited number of regional generation and transmission organizations which serve as wholesale power supply organizations to retail distribution systems organized, as today, under private, public, and co-operative ownership? Would such a reorganization permit the identification of prospects for improved efficiency in distribution, where major costs to the ultimate consumer arise? If found to be desirable, how could movement in the direction of such a system be achieved?

VI

Nuclear Energy

FROM A PUBLIC POLICY STANDPOINT, nuclear power is unique among the energy industries since from the beginning it has been the product of federal action arising out of a military effort launched during World War II. To a major extent, public policy issues regarding the peaceful applications of nuclear power have centered on the degree to which nuclear energy should be set free from federal government control and be permitted to enter the mainstream of the free enterprise system, and on the means by which such de-control could be accomplished. This is almost totally opposite to the situation prevailing in the other energy industries. These all began within the private sector, and public intervention has been the result of a subsequent realization of the need for various forms of regulation or assistance.

Although the direction of change in the field of nuclear energy has been unmistakably clear, the road from complete federal government ownership to free enterprise has not yet been completely traversed. In any event, the security and safety aspects of nuclear fuels will always require that a degree of government involvement be present. It will always be necessary to consider de-control within the framework of military security because the type of nuclear fuel used in U.S. power reactors could represent a starting point for the production of material that might be used in nuclear explosives. Moreover, from the start it has been recognized that radiation hazards associated with the use and disposal of nuclear materials give the government an important function to perform in the areas of health and safety. (The hazards created by fossil fuels, although of a different sort, have more recently been recognized as requiring governmental attention. See Chapter VIII.) Furthermore, the path that has been followed by the government to bring the peaceful application of nuclear energy to its present position, and the path that will be followed in the future, will leave a definite imprint on the institutional and economic structure of the nuclear power industry.

The Legislative Framework

The trend away from federal monopoly of the atom may be traced in three major pieces of legislation—the laws of 1946, 1954, and 1964. The original Atomic Energy Act of 1946 turned over the nation's total nuclear enterprise, created within the military establishment during the war, to the control of a civilian Atomic Energy Commision (AEC). But although civilian in membership, in its early years the Commission was concerned almost exclusively with the atom's military aspects (some of which have, however, yielded important civilian by-products). Secrecy and government monopoly of all nuclear activities were essential features of the 1946 law. Private enterprises were involved only as they functioned as contractors for operating federally owned facilities.

Major modifications occurred with the passage of the Atomic Energy Act of 1954. Among other changes, this law called for the declassification of much information that had been restricted, and permitted industrial participation and international co-operation in the development of peaceful uses of atomic energy.

Most significant of all was the cessation of the government's monopoly of reactor ownership. For the first time, private industry was permitted to own and operate nuclear reactors, including those for the generation of electricity. As it turned out, the government had much to do in order to bring about private participation, but legally the way was opened.

However, the 1954 Act still retained government ownership of all fissionable material. Private operators could obtain such material—the fuel needed for the reactors—only on lease from the federal government. And any fissionable material generated within a privately owned reactor was by law government property for which the government paid a "fair value" upon delivery.

In 1964, legislation permitting private ownership of fissionable material opened the way to the removal of this final vestige of federal monopoly. Full private ownership is to be reached in steps over a period of years. The AEC's continuing concern with national security and public safety is to be maintained through regulation and licensing.

From a beginning in which national security, public safety, and international relations were thought to require complete government ownership, an end point had been reached with the 1964 legislation in which private ownership of nuclear reactors and fissionable materials, suitably regulated, came to be regarded as a sufficient condition for guaranteeing the nation's vital interests.

The Competitive Status of Nuclear Power

The year 1966 is regarded as a landmark in the short history of the commercial nuclear power industry. For the first time, more than 50 per cent of the new electricity generating capacity ordered, or announced for future construction, consisted of nuclear power plants. These were commercial power plants, to be produced by private equipment manufacturers, with no apparent element of government subsidy. The era of competitive nuclear power appeared to have arrived.

But has a competitive status for nuclear energy really been achieved? Will nuclear energy dominate the new power capacity installed in the United States from now on?

It must be noted that no plant competitive with conventional power plants is yet in operation and that until real operating experience has been accumulated for both large nuclear and fossil fuel units, the factual basis for the atom's competitive status cannot be said to be firmly established. Moreover, the Atomic Energy Commission, although required by the Atomic Energy Act of 1954 to make a finding of "practical value" when commercial status for any reactor type has been achieved, has declined to make such a finding, even though petitioned to do so by various representatives of the coal industry in 1964 and again in 1966. Thus, the nuclear power plants now being constructed are still being licensed under the developmental, not the commercial, provisions of the 1954 Act. In denying the 1966 petition the Commission stated:

> Pending the completion of scaled-up plants, and the information to be obtained from their operation, the Commission remains of the view that there has not been sufficient demonstration of the cost of construction and operation of light water, nuclear electric plants to warrant making a statutory finding that any type of such facilities have been sufficiently developed to be of practical value within the meaning of section 102 of the Act.[1]

Although the finding of "practical value" may turn on narrowly legalistic questions, and the Commission may simply prefer the greater flexibility which is open to it under the developmental provisions, there are other grounds for raising questions concerning the true economic status of nuclear power. These point to areas, listed below, in which study is required in addition to the study continually under way within the Atomic Energy Commission. Outside, disinterested study is needed if only because the AEC is under statutory obligation to encourage the development of nuclear energy. Nevertheless, outside researchers should become

---

[1] U.S. Atomic Energy Commission, *Civilian Nuclear Power: The 1967 Supplement to the 1962 Report to the President* (Washington: Government Printing Office, February 1967), p. 56.

as familiar as possible with the AEC's own studies, and should employ
AEC analyses as inputs into their own studies.

1. Fuel Cycle Costs. Private ownership of nuclear fuel is only now in
the process of becoming a reality. The stages by which full private ownership is to be realized, and the remaining role of the federal government
during and after the various stages, is a maze which it is extremely difficult for the outsider to penetrate. The following quotes from the Commission's 1967 supplement to the 1962 report to the President on civilian
nuclear power illustrates the complexity:

> A key element of the Private Ownership Act was the provision for a
> transition period to private ownership. AEC may continue to distribute
> special nuclear materials by lease for power reactor use until January
> 1, 1971, and lessees may retain materials on lease until July 1, 1973. It
> is expected that reactor inventories gradually will be transferred to
> private ownership through toll enriching [an AEC service of enriching
> privately owned uranium for a fee] during the transition period, . . .
> [It should be noted that the majority of nuclear plants already an-
> nounced will not be obtaining their fuel before January 1, 1971, and
> will, therefore, not be affected by these transition provisions for ob-
> taining fuel.] . . . The transition is also being smoothed by a provision
> that guarantees a purchase price for plutonium or uranium-233 (if
> produced through the use of special nuclear material leased or sold by
> AEC) delivered through December 31, 1970. The guaranteed market
> for uranium-233, but not plutonium, may be extended by the AEC be-
> yond 1970. These prices have been set at the estimated fuel value of
> $10 per gram of fissile plutonium and $14 per gram of uranium-233. . . .
> The new law also authorized the AEC to provide toll enrichment
> services whereby raw materials normally will be purchased on the
> open market, rather than from the Government, and will then be pro-
> cessed to uranium hexafluoride in privately owned plants, enriched in
> the Government-owned gaseous diffusion plants, and further processed
> into fuel elements under private auspices. . . .
> . . . the AEC on December 23, 1966, established criteria under
> which the uranium enrichment services will be provided. The criteria
> provide that the AEC will offer to contract to supply separative work
> for periods up to 30 years at a unit charge to be announced, but sub-
> ject to a ceiling. [The ceiling price announced in December 1966 was
> $30 per kilogram unit of separative work. The effective price an-
> nounced in September 1967 was $26 for a kilogram unit of separative
> work. The "tails assay" to go with this price was 0.2 per cent. Natural
> uranium contains 0.7 per cent U^{235}; after the enrichment process, to
> whatever degree, the "tails" material will contain a reduced content of
> 0.2 per cent. Obviously, the greater the degree of enrichment in U^{235}
> desired, the greater the number of units of separative work that will
> be required. These prices are mentioned here for the sake of complete-
> ness, but their meaning, and the qualifications attached, cannot be
> appreciated without entering into a degree of technical detail beyond
> the scope of this report.]

In addition to the toll enriching criteria, a number of related AEC policies with respect to sale and lease of enriched uranium during the transition period have been announced in order to help the nuclear industry estimate the relative economics of sale, lease and toll enriching. In order to encourage the emergence of a viable domestic uranium market, the AEC will continue to base its enriched uranium sale or lease charge schedule on $8 per pound of U_3O_8 through June 30, 1973. . . .

Contracts offered by the AEC to implement the enrichment services criteria may be signed at any time, but the services may not commence before January 1, 1969. The contracts contain a provision permitting termination by the Government if comparable commercial uranium enriching services become available from another domestic source on reasonable terms and at reasonable prices within the AEC ceiling charge. This will permit the Government to be relieved, as may be judged appropriate, of strictly commercial commitments extending over the life of these contracts.

One possible source of a commercial enriching service could be the disposition of one or more of the three AEC gaseous diffusion plants to a privately or publicly financed corporation. The AEC is conducting an internal study on the feasibility and desirability of transfer of gaseous diffusion plants to private operation. The Commission is also discussing with the Atomic Industrial Forum the scope and ground rules for a proposed Forum study [now well under way] of such transfer.[2]

It is apparent that under government ownership during the transitional phase which ends on July 1, 1973, fuel cycle costs will reflect various prices fixed by the government, ranging from that for uranium concentrate (U_3O_8) through that for the enriched uranium which will continue to be available on a lease basis to private operators until January 1, 1971 (the annual charge of 4¾ per cent has been increased to 5½ per cent). These costs will also reflect the buy-back price of fissionable materials produced in privately owned reactors. Under the present plans, even after the transition to full private ownership has been achieved, the enrichment of privately owned uranium will continue to be performed at a fee (toll enrichment) in federally owned facilities. Thus, even after July 1, 1973, government pricing will continue to be a key link in the price chain.

However, the possibility is left open that enrichment services may be terminated by the government "if comparable commercial uranium enriching services become available from another domestic source on reasonable terms and at reasonable prices within the AEC ceiling charge." And the possibility is suggested that one or more of the three AEC-owned gaseous diffusion plants may be sold to a privately or publicly financed corporation, to provide enrichment services on a commercial

[2] *Ibid.*, pp. 53-54.

basis. Nor is it out of the question that completely new plants might be built.

In all this complexity, one fact appears clear: The AEC has in the past operated as a price administrator so far as fuel costs are concerned, and it is not completely relinquishing this role as long as it continues to determine the price at which toll enrichment is performed. In the past the accounting techniques for AEC costing have been reviewed inside the government by the General Accounting Office and the Bureau of the Budget, but for security reasons they have not been open to detailed outside study. Now the security curtain is beginning to be lifted [3] *and outside study is desirable for a public appraisal of the reasonableness of AEC pricing. With the discussion and study of the possibility of disposing of AEC-owned enrichment facilities, open study of the entire fuel-cycle cost/price structure should become possible. This is an important subject for early investigation because of its bearing on the real costs of nuclear power, now and in the future.*

2. Equipment Costs. The surge of orders for nuclear power plants over the past couple of years has been based upon equipment costs and performance guarantees offered by the equipment manufacturers. Until recently there had been only two successful bidders—General Electric and Westinghouse. Consequently, the competitive position of nuclear power as it is presently perceived may be said to mirror to a major extent the costs which these two companies have quoted.

Do the prices at which these two companies (and more recently others) have offered to construct nuclear power plants fully cover their own costs of development and production? Numerous reasons have been offered as to why a promotional pricing course might have been chosen, including the desire of each firm to establish its own position in what is assumed to be the growth sector of the future electric equipment business. It has also been reasoned that the sale of equipment, including the initial fuel cores, provides a guarantee of future business in the sale of reactor fuel cores. It is noteworthy that, unlike the situation in fossil-fuel fired plants, the major equipment manufacturers will also be in the business of selling fuel to electric power plants employing their type of reactor.

Some evidence that the equipment manufacturers had indeed been absorbing development costs is provided by the fact that quoted nuclear fuel plant costs reached their lowest point in 1966 in the announcement that TVA had gone nuclear in its Browns Ferry project. Since that time

---

[3] See, in particular, *AEC Gaseous Diffusion Plant Operations* (ORO-658) issued by the Commission in February 1968, after the present report was completed and submitted to the Office of Science and Technology.

quoted equipment costs have risen substantially (30 per cent or more), a rise greater than that of fossil fuel plants.

The basis on which nuclear equipment costs and performance guarantees have been (and will be) determined and the degree to which they conform to the realities of the situation is a subject requiring close study. The behavior of equipment purchasers may be presumed to reflect the realities of the market place as perceived by management in the electric utility business. But the market can be strongly affected by the promotional practices of equipment manufacturers. The factors motivating these practices may make quoted prices diverge from costs incurred or anticipated costs. Moreover, a bandwagon psychology, which produces temporary distortions in the functioning of the market place, can take hold. If the true competitive position of the alternative electric energy sources is being distorted, this is a matter of concern from the standpoint of efficiency of resource allocation. It is of concern also as a factor affecting the competitive position of the fossil fuels industries as well as the ability of other equipment suppliers to enter the nuclear power plant business and create a workably competitive situaion. To a degree the latter question presumably will come within the purview of a Justice Department–AEC sponsored study currently under way, focusing on the questions of present and prospective competition in the nuclear industry and on how the AEC and the Department of Justice can discharge their responsibilities to promote and maintain competition. Decisions concerning further study of equipment costing practices and their implications for the competitive status of nuclear power as an energy source should be made with due regard to the government-sponsored research now in progress.

3. Supporting Technical Programs of the AEC. As a continuing part of its program, the AEC supports technical programs in such fields as reactor safety, general reactor technology, and radioactive waste management. In the 1967 supplement to the 1962 report to the President, the AEC explains these activities as providing "the research, exploratory development, and advanced technology necessary for accomplishing specific reactor programs, for assuring the safety of reactor installations and for providing safe and efficient methods of handling the fission products and other radioactive waste products resulting from reactor operations." It goes on to say that "the success of these supporting technical programs continues to be a vital factor in the growth and acceptance of nuclear power." [4]

[4] U.S. Atomic Energy Commission, *Civilian Nuclear Power: The 1967 Supplement to the 1962 Report to the President, op. cit.,* p. 40.

These activities are of importance. To take one example, it is obvious that the engineering of safety features to minimize the consequences of accidents or to prevent their occurrence is a matter of urgency, particularly as reactor sizes become larger and the location of nuclear plants in proximity to load centers is pushed for economic reasons. It is equally clear that with the growth of nuclear power capacity, the disposal of highly radioactive wastes—which must be contained and isolated for hundreds of years—will become an ever-growing problem. Systems that have been used to date have been satisfactory on an interim basis, but do they, or known alternatives, provide adequate protection for the much larger quantities that will need to be contained over the indefinite future for which protective devices must serve?

Government expenditure on research and associated development and demonstration on an engineering scale in support of the emerging nuclear power industry is, in fact, an element in the costs of nuclear power but is not reflected in the costs that are quoted. *Study should be directed towards the measurement of these and other costs of nuclear power not covered in the market price. (Is the uncharged premium on governmental indemnity against the risks of a nuclear plant accident over and above the amount private insurers are willing to provide to be counted among such costs?) Without measurement of costs such as these, the real cost of nuclear power is not known.*

4. The Future Cost of Nuclear Raw Materials. If natural uranium costs rise in the future as a consequence of the growth of the nuclear industry, the cost of atomic power may rise above its current levels. This subject is dealt with on page 111.

5. The Real Costs of Nuclear Power. *The foregoing items are separable aspects of a single broad problem: the validity of quoted nuclear power costs for the reactors which are currently being ordered by power plants. The subjects can be addressed separately; but they should be regarded as aspects of the overall question of the true competitive status of reactors of current design. They call for open research, drawing heavily on AEC knowledge, on the structure of nuclear costs.*

.

One reviewer's reflections on the foregoing set of questions concerning the true competitive status of nuclear power seemed to us worthy of quotation at some length. We do so with his permission, from a letter dated November 1, 1967.

I doubt that two years ago anyone anywhere would have predicted the events that have since occurred. To be sure, many of us believed, from the time Jersey Central's Oyster Creek report was published in the spring of 1964, that utilities would begin to accept nuclear generation as economically competitive in some situations. However, we were not particularly surprised when it turned out that not a single nuclear power plant was ordered in that year. It seemed to us that, given the history of nuclear power construction ventures (adventures?) in the late '50's and the belief, which most of us held, that utilities are normally conservative in approaching investment decisions involving technical innovations, this was only to be expected.

The trend toward nuclear power did not really set in until just about exactly two years ago. Plans to build nuclear units representing about 5,000 mw were announced in the last months of 1965, and there seemed to be reason to hope that '66 would be a really good year. The results far surpassed our expectations: total nuclear power capacity ordered or announced for construction in 1966 reached some 20,000 mw, or more than half of all the capacity ordered or announced for construction that year. . . .

At the beginning of 1966, many knowledgeable observers concluded that the big push was over, at least for the time being. . . . We could not have been more wrong, as the history of the last ten months attests.

Meanwhile, industry and AEC estimates of the rate at which nuclear units are likely to be installed have approximately quadrupled in the last four to five years. Now estimates tend to focus on the range 150,000-175,000 mw for 1980. Whether these forecasts will be realized is, of course, far from clear. For my purposes here, the fact that they have been made and are widely accepted by organizations that are not wholly irresponsible is enough.

For spectators like myself, the grand question must be: What have been and will be the utilities' real motives in making so enormous·a commitment in so short a time? I think the studies proposed in your report ought somehow to come to grips with that problem. . . .

The big nuclear power units ordered beginning with Oyster Creek in '63 involve total capital investments amounting to, say, $7 billion to $8 billion. This makes no allowance for first fuel loadings. Altogether, these projects involve, directly or indirectly, at least fifty utility organizations, including municipals, co-ops and TVA as well as the private companies. One utility alone has ordered six units representing a total capacity of more than 4,800 mw; several have ordered more than one.

Can it not be taken for granted that behind each of these orders was the conclusion of the utilities concerned that nuclear power represented the best choice that could be made from the point of view of their own economic interests? And considering the number of orders, the number of utilities who have reached the same decision, and the fact that all but a few areas of the country are involved, is one not obliged to conclude that from the utilities' point of view the competitive advantage of nuclear power needs no further demonstration? . . .

Your report notes, appropriately, that the orders of the last two

years have been placed before anyone has had operating experience
with even a single commercial-sized unit—a fact which, one might
suppose, would normally discourage utilities from taking the plunge.
(One might add that the orders have been placed before anyone has
had start-to-finish construction experience or startup experience.)

It might also have noted that:

• All of the orders have been placed at a time when efforts to predict
fuel cycle costs for the next ten to fifteen years, let alone for the life of
a given plant, have been clouded by many uncertainties. (This is still
true and will be true for some years at least. . . .) While there are, of
course, uncertainties in estimates of the costs of all fuels, those concerning the nuclear fuel cycle are infinitely more numerous and
complex. Moreover, ownership of a nuclear plant involves its owner in
a formidable series of fuel management problems which he would automatically escape if he chose other fuels. It may also take him beyond
the energy conversion business and lead him into the market place as
a merchandiser of such products as plutonium and radioisotopes. . . .

• All of the orders have been placed at a time when none of the purchasers could say with much assurance when or how they would
emerge from the licensing process: what delays their projects would
be subjected to because of it, what added capital costs they would in-
cur, or indeed, whether their projects would be approved at all. . . .

• All of the orders have been placed at a time when any given utility
might well conclude that installing a nuclear unit would involve it in
short-term and, possibly, long-term public relations problems.

• Many of the orders have been placed two or even three years ahead
of the time construction was expected to begin, and have involved the
utilities in commitments that they could certainly have delayed if they
had chosen to build conventional units.

• Many of the orders have been placed at a time when nuclear power
equipment costs have been rising sharply. . . . the increase has been
somewhat steeper than the rise in comparable costs for conventional
plants.

It would be easy to cite other factors which normally would be considered deterrents. Together, they make an impressive list. . . .

If I were asked whether I thought that all of the pro-nuclear power
decisions have been based on thorough studies illuminated by enlight-
ened self-interest, I could not say yes. I think there may even be
something in the idea that one sometimes encounters these days that
what we are seeing reflects the general adoption of a fashion. But on
the whole this is surely superficial. Certainly some utilities have been
swayed by the fact that the industry's acknowledged leaders . . . have
made heavy nuclear power commitments. And certainly some plants
have been ordered almost on the spur of the moment. However, I do
not think this implies that the decisions have not been made for very
sound reasons from the individual utilities' points of view. To me, this
suggests that the potential economic advantages to the utilities are so

clear as to be almost self-evident. And I think we will not identify those advantages just by questioning whether nuclear electric generating costs are indeed competitive, as they are said to be. Only a broader inquiry will let us do that.

All of this, I suspect, must have some important implications for analysts and policy planners. In particular, isn't it possible that the disconcertingly sudden shift to nuclear power . . . is typical of what we must expect in the future of the energy industry as a whole? . . .

To return to the point I have made earlier—a broad study of the shift to nuclear power is desirable if not essential, and it should be done on a broader base than the one suggested, and not merely in terms of questions with negative implications.

The study I have in mind would indeed ask whether the economics of the nuclear fuel cycle economics are in fact so very attractive. It would include asking whether the supposed advantages depend significantly on covert subsidies. And doubtless, it would look critically at capital costs, for both nuclear and fossil-fueled units.

However, it would also ask whether or to what extent cost penalties are imposed by AEC policies that are not primarily derived from economic considerations. And it would try to determine whether any penalties of that kind, or other factors which have an adverse effect on nuclear power economics, will be likely to evaporate, and when.

Thus, to give just one example, it would not ask only whether the AEC's uranium enrichment charge, as recently established, is lower than is justifiable from the point of view of cost recovery. It would also ask whether the charge has not been set at an artificially high level. And if it found that this has indeed happened, it would try to assess the chances of a reduction.

Federally Financed Research and Development

In its 1962 report to the President on *Civilian Nuclear Power*, the AEC estimated that $1.275 billion had been expended by the government to date on the civilian power program, and that industry had expended approximately $0.5 billion of its own funds, mostly for plant and equipment.[5] Currently, according to its *1967 Financial Report*,[6] the AEC is spending more than $500 million per year on reactor development, of which between $150 million and $250 million is assignable to commercial nuclear power. What is the justification for continued expenditure of such large sums for research and development now that commercial nuclear power has apparently been achieved?

The main reason offered is that the class of reactors that are now thought to be economically competitive with fossil-fuel fired power plants

[5] U.S. Atomic Energy Commission, *Civilian Nuclear Power: A Report to the President, 1962* (Washington: Government Printing Office, November 1962), p. 8.

[6] U.S. Atomic Energy Commission, *1967 Financial Report* (Washington: Government Printing Office, 1967), p. 5.

are inefficient in their use of nuclear raw materials. Whereas a light water reactor of current design is capable of converting only a slight percentage of natural uranium into fissionable plutonium, other more advanced designs are capable of achieving conversion rates of better than 50 per cent. The goal of achieving more efficient resource utilization is judged by the AEC to be important, and it has consequently been pursuing an R&D program directed towards this end.

Oddly enough, the unexpectedly rapid adoption by electric utilities of the light water reactor appears to make attainment of this goal more urgent than was formerly thought, because projected uranium usage by such reactors now seems to pose an earlier threat to the depletion of known and estimated domestic uranium resources than had been expected. By the same token, the unexpected success in reducing costs of the light water reactor may have made it more difficult to achieve the economic target for the more efficient raw material converters; greater efficiency in raw material usage must be achieved together with overall power generating costs below those expected just a few years ago.

The drive to achieve reactors that are more efficient in their use of nuclear raw materials gives rise to a number of questions which require study. *The most basic question is whether the federal government's concern with the adequacy and cost of future energy supplies justifies any further research at this time, and, if so, how much should be devoted to improvement in the utilization of nuclear fuels. Uranium is but one of several fuels that are used to generate electricity. The main one used today is coal. Decisions concerning R&D that are motivated by a concern for the efficiency of the utilization of uranium, or any other fuel resource, cannot be properly evaluated except in the context of a consideration of the supply and demand outlook of all fuels used in generating electric energy. This question is posed in Chapter VIII.*

At this point, however, the more basic question will be bypassed, and the nuclear R&D issue will be addressed solely within the confines of the technological choices that are open within the atomic energy field. Numerous alternative reactor designs are available for further development. The Commission appears to be pursuing an R&D strategy in which: (1) the heaviest emphasis is placed on the fast breeder, which produces the highest gains in converting natural uranium into fissionable plutonium; and (2) provision is also made for R&D in so-called "advanced converters" in which uranium is utilized more efficiently than in the present generation of light water reactors, but not as efficiently as in the fast breeder, and which pose less difficult developmental problems than the fast breeder. The advanced converter is thought of as bridging the period between the present low-yield nuclear reactor and the ultimate high-yield nuclear breeder reactor.

The AEC's R&D strategy, although plausible, appears to us to rest essentially, although not wholly, upon a raw-materials calculus, and the question which must be examined is whether it is optimal in terms of other criteria. The main alternative criterion against which the strategy must be assessed is the economic one of cost effectiveness. If reactor development were to be approached in terms of maximizing nuclear power cost reductions relative to R&D expenditures, what reactor design choices would be made over the relevant planning period? It may be that the raw materials efficiency criterion and the cost effectiveness criterion are not in conflict; only an open study of the R&D implications of the two, by an impartial study group, can determine whether they do or do not yield different R&D strategies. The social choice between the strategies, or some amalgam of the two, can then be made on an informed basis. Such a study, which is urgently needed, should rely heavily on analyses and other background information prepared in the AEC.[7]

Natural Resources of Uranium and Thorium

Because of the characteristics of its fuels, nuclear energy is unlike any other energy source that the world has yet known. The energy concentration within these fuels is so vast as to be almost incomprehensible by conventional standards. One pound of nuclear fuel is the energy equivalent of 2 to 3 million pounds of coal.

But nature does not yield this bounty easily. Despite the vast scientific and technological endeavor which has been invested in harnessing nuclear power in a controlled reaction, not much more than the small percentage of uranium (1 part out of 140) that is fissionable in its natural state (U^{235}) becomes a reactor fuel with present-day technology. Potentially, however, the plentiful isotope of U^{238} can be converted into plutonium, which is fissionable, and thorium can be converted into U^{233} which is also fissionable. Advanced converters, and ultimately breeder reactors, are the devices by which such transformations will take place.

Were breeding to be achieved, the resources of uranium and thorium would dwarf those of all the conventional energy resources by a tremendous margin. Indeed, it is likely that the achievement of economical breeding would remove any danger of energy resource depletion for as far into the future as anyone might reasonably want to look. It is easy to see, therefore, why the attainment of breeding should appear so enticing a technological goal. Certainly, this must be a long-term objective of nuclear reactor technology.

[7] See, for example, the studies referred to in U.S. Atomic Energy Commission, *Civilian Nuclear Power: The 1967 Supplement to the 1962 Report to the President, op. cit.*, p. 4.

In the meantime there also may be a problem of natural resource adequacy, in terms of sufficiency of uranium to fuel the rapidly growing nuclear power industry. The cry of resource scarcity has been raised not only by those who favor the idea of having the government invest huge sums in an attempt to hasten the achievement of high-gain breeder reactors, but also by those who would like to discourage utilities from deciding to construct nuclear plants at the present time and in the near future.

According to the AEC,[8] known and estimated *domestic* resources of uranium at prices less than $10 per pound of uranium oxide (U_3O_8) are adequate to meet light water reactor requirements for about the next twenty-five years. Although current exploratory activity is at a high level, this earlier evaluation suggests numerous near-term questions which require early study.

1. *How valid are the uranium resource estimates? What can be learned through geological inference concerning domestic uranium supplies that would be forthcoming at a schedule of prices of uranium oxide (U_3O_8) higher than the $10 per pound upon which current figures on known and estimated domestic resources are based? Attempts to estimate such a supply schedule, along with similar supply schedules for other fuels, is identified as an urgent need in Chapter VIII.*

2. *What techniques, if any, should the government employ to step up uranium exploration in the United States? Or should the search be left wholly, or mainly, to private initiative now that the nuclear power industry has started a period of rapid growth, nuclear fuel has been almost completely de-controlled by the federal government, and the tempo of exploration by private industry has quickened?*

3. *Should the embargo on the importation of uranium be lifted? The embargo was introduced to permit the domestic industry to survive during the period in which the uranium market was in the doldrums. Is there reason for its continuation in the changed situation?*

4. *Should the growth of light water nuclear power capacity be slowed down so that capacity will not outrun fuel supply? If so, what should be the criteria for adopting such a policy, and what are the techniques by which it might be implemented?*

5. *To what extent should the substantial AEC uranium inventory, in excess of military needs, begin to be sold on the open market, and at what price?*

[8] *Ibid.*, p. 14.

Nuclear Energy and Foreign Policy

From its earliest days, atomic power has been enmeshed in foreign policy considerations. At the beginning, when the possibilities of regulation and inventory control were not as well understood as they are today, the United States, in a dramatic move, proposed before the United Nations the creation of an international authority for developing the peaceful use of atomic power and for preventing the diversion of nuclear fuels to weapons use. The plan was not adopted. Other initiatives were adopted at later dates, including the Atoms for Peace plan first offered by President Eisenhower in a speech before the United Nations in December 1953. The plan was premature in its nuclear power aspirations, and unfortunately it aroused false hopes regarding the possibilities for nuclear power power plants as an aid to industrialization.

The danger of the proliferation of nuclear weapons as a result of the international spread of commercial power plants has never ceased to be a matter of concern. At times, it has even been proposed that peaceful applications should not be pursued on the grounds that the benefits to be gained would be far outweighed by the risks entailed by weapons proliferation. The argument is advanced today, particularly in connection with plans for the constructive application of nuclear explosives.

A comprehensive appraisal of the foreign policy aspects of nuclear energy from a present-day U.S. perspective is an important matter for study. It should address such questions as the following:

1. *Can the U.S. reactor development program be made responsive to the energy needs of other countries? What types of reactors would best meet the needs of those countries which can accommodate large-scale installations, but which wish to avoid the balance-of-payment and fuel-dependency problems that might be associated with the need for acquiring enriched uranium from the United States? What types of reactors would best meet the needs of those countries, such as India, which are well endowed with thorium, but not with uranium? Can a greater amount of research be usefully devoted to small-scale reactors that could meet the electricity needs of the bulk of the world's nations which cannot absorb the gigantic nuclear power plants resulting from current and projected U.S. technological programs? What would be the costs and benefits of such a redirection of program?*

2. *Does the nuclear power-desalting-industrial-agricultural complex concept recently explored at the Oak Ridge National Laboratory hold real promise for the less developed world? Can it be scaled to meet the*

level of power and water needs in the less developed countries, and still be economically feasible? Is desalting technology far enough advanced compared with nuclear power technology to render this combination realistic? If such a complex has merit, would power from sources other than atomic energy be preferable in some instances?

VII

Shale Oil

IN RECENT DECADES, whenever there was reason to question the future adequacy of petroleum supplies, the prospect of oil from shale—hundreds of billions of barrels—has served to alleviate fears. And for good reason. Within the three states of Colorado, Utah, and Wyoming, underlying an area of only 16,000 square miles, there are deposits of oil shale estimated to contain the equivalent of close to three thousand billion barrels of oil. Some mineral lease-size plots (5,120 acres) are believed to contain as much as 18 billion barrels, equal to more than half the country's proved reserves of crude oil.

The deposits—which consist of marls containing the organic material kerogen from which oil is derived through retorting at high temperatures—differ widely in quality, both horizontally and vertically. In some parts, thickness runs up to 2,000 feet; in others, it measures as little as 15 or 10 feet. High-grade deposits may be often interbedded with low-grade deposits, or separate rich zones may appear at different depths in the same location.

Nonetheless, one may generalize that despite uncertainties regarding grade and thickness, the exploration needed to acquire adequate knowledge for commercial operation is hardly of the magnitude of that required for many other minerals, including petroleum. The U.S. Geological Survey's estimate that the area contains 80 billion barrels of "known resources, recoverable under present conditions," [1] may thus be accepted as of relatively high reliability. This volume is contained in favorable formations defined as yielding 30-35 gallons per ton, in seams more than 25 feet thick and extending to a depth of not more than 1,000 feet, with 50 per cent of the shale in place recoverable and 50 per cent of

[1] U.S. Geological Survey Circular 523, *Organic-Rich Shale of the United States and World Land Areas*, by Donald C. Duncan and Vernon E. Swanson (Washington: 1965), p. 13.

the energy in the shale recoverable as commercial output. Beyond are vast quantities in locations that possess less attractive characteristics.

However, even those deposits defined as "recoverable by demonstrated mining and retorting methods" are not yet known to be economically recoverable in competition with crude oil or other sources of energy. Conflicting estimates, hints, and surmise cover a range of costs. There is a fairly widely held view that costs of producing shale oil are "close to competitiveness." This vague opinion defines our present state of knowledge—or lack thereof. Ongoing research within the federal government is, however, taking new aim at the probable costs of production, the relation of estimated costs to crude oil costs, and the implications of the relationship of the two for estimates of the timing of shale oil's entry into commercial production. The prospects for the development of this vast natural resource involve far-reaching decisions by the federal government, especially with regard to tax policy and leasing policy. The process of development will be much complicated by problems of waste disposal, water requirements, community development, transportation needs, and the fate of other minerals found in association with oil shale.

Federal and Private Ownership

Basic to an understanding of shale oil problems is the fact that the federal government owns at least 72 per cent of the country's oil shale acreage, and, as this area contains a high proportion of the richer acreage, close to 80 per cent of the estimated oil in place. Because this calculation omits acreage to which title is currently in dispute, the percentages cited may be presumed to be conservative.

The federal government's role as owner is reinforced by an Executive Order of 1930 that withdrew all federally owned oil shale land from leasing. Private enterprise has thus been unable to gain further access to this land, some of which consists of the richest and most desirable acreage in the region. Much has been made of this circumstance. The federal government's predominant ownership role and its refusal hitherto to permit exploration or development of its land for shale oil production have frequently been represented as the principal block to the emergence of a profitable shale oil industry. Whether such a view is prompted by hopes of developing an industry or by the desire to control valuable pieces of mineral land, it stands in urgent need of dispassionate scrutiny. It must be assessed, for example, against the fact that in Colorado alone private companies are known to hold 168,000 acres of oil shale land; or that all land privately owned is estimated to contain 150 billion to 200 billion barrels of oil equivalent. This does not mean that all of this land is easily exploited. Shape, size, or other characteristics may make commercial working costly or otherwise inadvisable. But more than assertion seems called for in support of the thesis that the insufficiency of prime oil shale land in private hands prevents the development of the industry.

Moreover, the estimated holdings in the hands of individual companies are quite large. For example, the largest on record, held by Standard Oil of California, have been estimated at 9.2 billion barrels of oil in place, on 50,000 acres of land. This is followed by Union Oil's holdings of 40,000 acres, estimated to be underlain by 8.8 billion barrels of shale oil. Even the smallest parcels of acreage held by oil companies are estimated to contain deposits of a magnitude that, were they to consist of crude petroleum, would earn them the appellation of "giant fields."

Not only are large acreages in private hands, but there is also much trading. A recent compilation showed that close to 100,000 acres changed hands between 1963 and 1965, at prices ranging as high as $2,000 per acre.

Statements that federal ownership combined with absence of leasing is a major obstacle to the emergence of a domestic shale oil industry thus must be subjected to careful review, especially as there are other reasons to explain the potential developers' reluctance to invest and develop. Federal, as compared to private, holdings may be enormous, but private holdings appear sufficiently large to support commercial shale operations. If so, what factors have blocked development?

In the first place, the investment in a commercially feasible operation, commonly thought to run around 50,000 barrels of oil per day, has been put at between $100 million and $150 million, all of which must be spent before the first drop leaves the retort. To pioneer such a venture would be a major undertaking even if the process were fully proved out. It represents an especially serious risk at an early stage when reliance must be placed on pilot plants. Moreover, the operation is limited to an area that is relatively far from markets and lacks ready means of transportation. Also, it must compete with an established oil industry endowed with vast resources in all phases of production, transportation, and distribution.

Second, given the size of the investment and the difficulties that the use of other than established distribution and marketing channels would encounter, the prospects are unlikely to attract many investors outside the oil industry. (The current attempt of the Colony Development Company to establish and put into operation by 1970 a 58,000-barrel plant drew initial support from both a mining and a major oil company, but the status of the respective commitments appears to be in doubt.) It is sometimes thought that investment in shale oil has been deterred by the tens of billions of dollars invested by the oil industry in this country and abroad in oil-producing properties and in transportation and refining facilities associated with them, which may be affected in a variety of ways by competition from shale oil. The low rate of investment in shale R&D has been cited in support of this thesis. But, as pointed out by one of our reviewers:

> The oil industry considered as a whole—i.e., as a monopolist—might very well want to limit or stop shale oil development. But the industry has never been able to act as a monopolist. In selling natural gas, each company has been forced to consider its own profit-seeking interest regardless of what it would do to the industry as a whole. The same goes for shale oil. If therefore no company has been willing to start commercial development of shale, this is not to be attributed to the fear of making past investment valueless, for a company going into shale development would bear only a tiny part of the industry loss and make all the gain. Hence the failure of private industry to make the move must be ascribed, up to this moment at least, to a conviction on everybody's part that given current prices and costs, shale development is not yet worthwhile.

Nonetheless, there might be a grain of truth in the contrary assertion, for while the initial entrant would "make all the gain," in the longer run the opening of the oil shale lands to commercial development could have far-reaching consequences for the production and pricing arrangements of all oil producers. The fact that a member of the industry has played a leading role in setting up the first tar sands plant in Canada is not a helpful guide, since that operation is taking place in a strictly managed phasing-in process.

Third, potential developers would like to see a number of elements associated with shale oil production resolved before they invest in it. Some of these questions can be answered through privately financed research and development; others, discussed below, must wait for governmental policy decisions.

None of these elements would be likely to prevent the development of a shale oil industry if it were not for a fourth one—*viz.*, the inferior competitive standing of shale vis-à-vis crude oil—and to this we now turn.

Cost Analysis and Interindustry Cost Comparisons

Among the important issues awaiting clarification is the matter of costs. This comprises meaningful comparison between prospective costs of different methods of production and between shale oil and conventional oil production, both for average and marginal costs, as well as attention to external costs, which in the case of shale might be substantial, given the heavy disposal load of spent rock and the social policies relating to environment.

At the present time, little can be said about comparative costs of different possible methods of recovering shale oil. The two basic methods are: (1) mining the rock with surface processing; and (2) *in situ* recovery underground, possibly including novel ways, such as nuclear explosions, of fracturing rock formations. With respect to *in situ* recovery, technology has not advanced sufficiently to yield credible cost estimates; it would probably take an extended process of research and development to give any substance whatever to estimates that have been proffered. With respect to mining and retorting methods, a considerable amount of attention was given to cost estimation prior to 1951. A new wave of interest is causing earlier estimates to be updated. Generally, these estimates lead back to experimental efforts carried on by the Bureau of Mines and to data made available by engineering firms that have specialized in this field.

At the center of interest in such estimates are the comparisons that may be made between them and the costs of crude oil, since in this relationship lies the key to commercial shale oil production. The cost structure of shale oil production differs radically from that of crude petroleum in the following ways: (1) The finding costs which figure substantially in petroleum are minimal in the case of shale. (2) The material lifted to the surface is rock rather than oil, and thus involves mining technology and problems. (3) A commercially acceptable technology is required to retort the rock and recover the liquid. (4) Substantial costs are involved in the disposal of the residual mineral material. There is also an additional processing stage at which the product's viscosity must be sufficiently reduced to permit pipelining. In different degrees uncertainties about eventual commercial costs relate to all phases of this process.

Work now in progress in the federal government—in the Bureau of Mines and by consultants in the Bureau of the Budget—is designed to update the analysis and to use more sophisticated methods to arrive at an understanding of the possible shape of the cost curve of shale oil production over time, and the timing of the entry of shale oil that would result from the assumed conditions. These efforts are to be welcomed. They should open up a long-needed discussion on the basis of specifically stated, and thus rationally debatable, assumptions concerning the conditions under which shale oil might enter the fuel market.

The most critical points to be examined are the marginal costs of shale oil in comparison with the short-run and long-run marginal costs of crude oil. (Eventually, of course, cost comparisons must reach across the whole spectrum of energy sources.) In such a comparison, a number of points at once present themselves, bearing upon the incentives to engage in development of shale oil resources.

1. Given (a) the surplus producing capacity of the crude oil industry and (b) the opportunities for secondary recovery from existing reservoirs, the relatively low marginal costs of increasing the short-run production of crude oil stands as a barrier to the development of shale oil production.

2. This barrier could, however, be breached by the long-run finding costs for new crude oil. Business judgment within the crude oil industry with respect to the profitability of further investment in exploratory activity will have a strong bearing upon the timing of the development of the shale oil industry.

3. Much could depend upon future trends in state regulatory policy. Restricting the output of crude oil, and thereby supporting its price, would in principle favor the early appearance of commercial shale oil. But the existence of discretionary state powers introduces an element of uncertainty, in that state regulatory authority could be turned to protecting oil producers against the inroads of shale oil.

4. Another factor to be weighed in the balance is the future of oil import policy. If present policy could be counted upon to continue, it would reinforce state regulation of oil in improving the prospects for shale oil. Any doubts about the continuance of the policy would have a deterrent effect. In an administrative context, oil shale producers will want to know if shale oil will be counted as refinery throughput and if they will thus be entitled to the receipt of "import tickets" (cf. Chapter II). Cost calculations would be affected thereby.

The points raised above lead to the following thoughts on lines of cost analysis relevant to shale oil policy:

1. *A better understanding of the variables affecting the cost of producing shale oil is called for, insofar as this can be done prior to the additional development activities that will have to be carried out.*

2. *Special attention must be given to costs lying outside the production process, in the fields of spent rock disposal, water pollution, and other environmental damage. Such costs, formerly often accepted as external social costs, are now likely to be placed as a charge upon private enterprise.*

3. *The quality of any judgment of the outlook for shale oil is as much dependent on appropriate cost analysis for crude oil as for shale oil. (The subject was dealt with in Chapter II.) Significant phases of oil costs involved are: (a) short-run and medium-run supply curves; (b) discovery costs for new reserves and development costs for old reserves; (c) the cost structure, now and in the future, resulting from*

state regulatory policies, reinforced by federal control; and (d) fiscal treatment.

4. *The quantities of water available for shale oil production are a critical limiting factor in the development of a shale oil industry.*

5. *Locational advantages and disadvantages need to be considered; for what eventually counts is not the cost at the retort but the cost at the point of consumption.*

The comparative cost information that becomes available will have an important bearing on the investment decisions of private companies. For purposes of public policy it has an importance of a different sort.

Given the effect of marginal costs in the short run, the existence of excess capacity in crude oil, and the leeway for lowering crude oil costs, private investment might be forthcoming wherever it originates, even when leases on federal land become available, only on terms that are equivalent to a subsidy. The question would then arise of whether the federal government should assume responsibilities reaching beyond the provision of a favorable investment climate.

Nature and timing of an emerging shale oil industry will no doubt be affected by policies yet to be formulated by the federal government. These will involve taxation, leasing of public lands, and support of research and development. Indeed, it is often alleged that the failure of the government to adopt firm policies on these matters is the greatest obstacle to the emergence of a shale oil industry, even on privately owned lands. Regardless of the validity of this assertion, policies will have to be set and their early appraisal will be helpful in determining both their urgency and their implications.

Taxation Policy

While not entering into the calculation of costs of production per barrel, *tax treatment* will closely affect profitability of the shale oil undertaking and obviously affect the decision to produce or not to produce. Under present law, the depletion allowance is 15 per cent of the value of the crushed shale. Even if this value were to be based upon cumulative cost to that stage of production, it would be greatly below the value of the oil as it emerges at the end of the process. The present depletion allowance, therefore, would affect net revenue far less than in the case of crude oil.

Many opinions have been advanced on this score. Some have argued that a 27½ per cent depletion allowance on the value of oil recovered, as is accorded crude oil, would make shale oil production commercially feasible today. Industry statements have been made that such an allowance would be equivalent to a 50-cents savings per barrel. Others have argued that, whatever the rationale of a depletion allowance for crude petroleum, it has no relevance to shale oil because there is no exploration cost or risk and no reason to allow for writing off a capital value greater than the depreciable cost of investment in the properties. A question of principle, therefore, is involved; *viz.*, whether or not anything in the form of special tax treatment or subsidy is properly applicable to the circumstances of shale oil production.

The benefits conveyed by percentage depletion to shale oil are qualified, as in the case of other minerals, by the statutory limitation that, regardless of gross revenue, the allowance may not exceed 50 per cent of net income before depletion and taxes. Thus, it cannot convert a net loss into a profit situation and therefore puts brakes on the extent to which profits can be raised. But within these limitations, percentage depletion could give shale oil the key to commercial success. For instance, Union Oil officials have occasionally been quoted to the effect that a $27\frac{1}{2}$ per cent depletion allowance (presumably on the liquid) would make their process profitable. Whatever the numerical value that would constitute the threshold, this would scarcely represent a rational approach in the establishment of a new industry. It does, however, serve to raise important questions for research, and above all, *whether equity requires that all sources of oil receive the same percentage depletion. A subsidiary question is whether all production processes of shale oil should receive the same fiscal treatment.*

What is called for is:

- *An analysis of the rationale of the depletion allowance for shale oil production as a whole, and for each of the processes.*
- *A calculation of the effects on profitability of different levels of a depletion allowance, including a zero level.*
- *An analysis of the implications of various levels for resource allocation, including the effect on competing energy sources.*

Leasing Policy

A second public policy issue of major importance to a shale oil industry is federal *leasing*. Since much of the valuable acreage is in federal ownership, sooner or later a leasing policy will be called for; and the sooner it is established and announced, even if effectiveness is postponed until some time in the future, the sooner other steps can be taken toward determining the future of the industry.

The importance of leasing rules lies both in the effect on cost and on the degree to which they will promote or impede efficient exploitation of the deposits. The choice between bonus bidding, royalties, and profit-sharing, or possible combinations of these methods, their respective levels, the size of the lease tract, and the conditions surrounding it (such as requirements covering effect upon environment and promotion of research and development) will all be reflected in profit calculations, and therefore figure in investment and production decisions.

Among important aspects of leasing policy calling for research are the following:

1. Size of Lease. The size problem arises out of the great qualitative diversity of shale land. If leases were to be of the size designated under the Mineral Leasing Act—5,120 acres—the shale oil content would in some locations run into billions of barrels. Tracts of such size raise special problems. If technology makes shale oil profitable, the lease bonus might limit bids to a very few companies, with adverse effects on competition. Given the initial uncertainties, errors of judgment in bidding would produce corresponding outsized financial penalties or windfalls.

Moreover, since there is a problem of phasing shale oil into total oil production, the rapid opening up of vast deposits might lead to serious market disturbance. Although the large sums of capital required per unit of output might mitigate this danger, it may become important to establish an "optimum annual supply" of federal oil shale lands. (The Canadian policy of permitting entry of tar sand oil on the basis of setting apart a quota of the estimated total oil market is of interest in this respect.)

The question thus arises of whether it would not be more reasonable to determine lease size on the basis, not of acreage, but of estimated shale oil content or recoverability and its relation to the requirements of an efficient operation and the associated investment, varying the size of the parcel accordingly. It has been suggested, for instance, that in the rich shale portions of Colorado, a parcel of 200 acres would be sufficient to permit profitable operation of what is commonly considered a reasonably-sized operation of 50,000 barrels per day. These considerations suggest, in turn, that tracts should be thoroughly core-drilled—or otherwise explored—before being offered for lease. Much of the above seems to have already been carried out, and accounts have been published by the Department of the Interior. Another issue raised by size limitations is the effect on profitability of pipelines that would carry the oil from the producing area, and that would need a multiple of such parcels if they were to be run at low unit costs.

Research is, therefore, recommended on the relationship of size of lease to nature of deposits, effect on competition, effect on financial capability of lessee, effect on competing energy sources, especially crude oil, and the economics of processing.

2. Leaseholder. If leases are kept reasonably small, though large enough to permit efficient operation, the limitation of no more than one lease per holder loses relevance to shale oil operations, as long as competitive bidding seems to be taking place. In fact, lifting this restriction might help rather than hinder competition.

Research would be useful which analyzed the working of this "one lease per holder" restriction in mineral industries most comparable in their characteristics to shale oil, and which speculated on various ways the restriction might be modified.

3. Lease Terms. Except for the feature discussed in the next paragraph, oil shale lands do not present any particularly novel features, with regard to lease terms, that distinguish them from other mineral lands. Auctions vs. sealed bids, royalty vs. profit-sharing schemes, as well as the levels of the respective charges, are issues that are not affected by the nature of the resource in question. The fact that a royalty payment becomes part of marginal costs and, therefore, leads to a termination of production earlier than would have been the case without such a payment (and thus is wasteful) is true not just for shale oil. Any remedies (such as waiving royalty payments at a late stage of exploitation) would be of general applicability.

What is most in need of analysis is the effect of different levels and combinations of lease bonuses and royalty or profit-sharing payments on the cost of shale oil production and its competitive position. The degree and kinds of uncertainty present in shale oil production might play a large part here, though these are bound to diminish as the industry gets under way. To illustrate the types of considerations, it has been suggested that the circumstances of shale oil might call for a high bonus payment (high in the sense that the government would want to set a high refusal price), a high annual rental to encourage production (with rental related to bonus rather than acreage, given the wide diversity of shale land quality), and a low royalty, since the element of risk-sharing is not prominent. Also, given the possibility of insufficient competition among bidders, there might be the need of a refusal price to protect the public interest.

Hypotheses directed at the appropriate levels of combinations of payments, and their effect upon the competitive position of shale oil, need to be carefully analyzed. For example, there is a question whether the same level and mix should apply to deposits exploited with different technologies. While standard room-and-pillar mining might call for one approach, large open-pit mining might call for a different one, and the rules applied to *in situ* recovery might again be different. *Thought should also be given to methods of preserving maximum flexibility over time, compatible with enough stability to provide potential producers with a reasonable basis in their planning operations. Policies in the early stage of the industry might differ from those appropriate to a mature industry.*

4. Scope of Lease. It has been suggested that, apart from its traditional functions, the lease is a proper instrument to provide for: (a) the carrying out of needed research and development; and (b) protection of the capacity of the land surface for uses other than mining and, more generally, of environmental factors that are affected by mining. The subsequent discussion of R&D is addressed to the use of leases in that connection.

As for terms that protect the multipurpose character of the associated natural elements, there are two important points to investigate: (a) To the extent that shale oil operations would be taxed with obligations that are not imposed upon competing resources and hence would make shale oil show up less favorably, is it in the public interest to cause such an effect? (b) What precisely are the associated costs that should be allocated to shale oil production? Should a lease specify restoration of surface land to a predetermined level and carry enforceable standards of air and water quality? Should it additionally provide for some kind of contribution to community development, or compensation for burdens placed on roads or other transport media, etc.? Society could as easily err to excess in the case of an industry now coming into view as it has erred by omission in the past. Proper demarcation will be needed before leases are written and standards frozen into practices.

Research and Development

Several factors have been responsible for making the promotion of the research and development of shale oil production a focus of attention as well as of controversy. One has been that an industry has failed to emerge, and that the results of private experimenting, understandably, have not been disclosed in any meaningful way. A second has been the relatively small amount of money that industry has devoted to shale oil research: the current effort at the Rifle, Colorado, plant of the U.S. Bureau of Mines, funded by half a dozen major oil companies at $5 million for a three-year period, can hardly be considered a major push. A third has been the intermittent nature of federal involvement. And a fourth has been the hope that a production technique might be developed that would eliminate the expensive and waste-generating mining stage. Perhaps the dogged persistence of local interests, centered in the state of Colorado, should be added as a fifth element, for it has often supplied the impetus for at least a review of the situation, and has led to the establishment of special shale oil programs in local educational institutions such as the Denver Research Institute and the Colorado School of Mines.

At issue are principally two problems: (1) What instrumentalities should be used in the conduct of research? (2) How much R&D effort should be devoted to the promotion of shale oil production? The potential carriers of R&D range from the federal government as a performer to private industry as an independent operator or as a contractor with the federal government.

The Bureau of Mines has for many years engaged in shale oil research, built experimental plants, surveyed and measured shale land, and published its findings. But the amounts of effort and funds so used have been very small; and discoveries of large new petroleum reserves in the United States and abroad have slowed the pace of research. The last time special funds were made available was in 1944 when, out of wartime concern for the adequacy of domestic petroleum supplies, Congress passed the Synthetic Liquid Fuels Act, under which the Bureau of Mines conducted research through 1955, including establishment of the experiment station at Rifle, Colorado.

Because of the slow and intermittent pace of research, thought has been given to setting up incentives toward increasing R&D expenditures and directing them into channels considered most in need of funds. One of the alternatives is direct contracting between government and research agencies, as is common in other fields of government-funded R&D work. Such contracts could be confined to specified phases of mining or processing, with the results made available under licensing to potential producers.

Another alternative would capitalize on the dominant position of the federal government as shale landowner and use leases as vehicles for promoting desirable research. *The "R&D lease" which has turned up as a development tool raises a number of questions. One of them is the degree to which considerations of equity would allow such contractors to receive favorable treatment, both vis-à-vis companies that do not contract for R&D and vis-à-vis companies that have in the past funded experimental shale oil work. Other problems are the definition of "success" in R&D work, the size of such leases, and the transition from an R&D to a full-scale commercial venture lease.*

A second, and wholly separate, problem is the determination of the relative magnitude of government funds to be devoted to shale oil R&D.

This is a matter that concerns not only shale oil, but all energy sources. Chapter VIII contains a tentative list of some criteria that might be considered in assessing the need for government R&D funding. In the case of shale oil, one is inclined to establish the following ratings: (1) certainty of success as high; (2) time horizon of payoff as reasonably close; (3) diffusion of expected benefits as very great; (4) magnitude of privately-financed R&D as very small, (5) external effects as many and sizable; and (6) size of investment required as large. *The criteria available do not yield an unequivocal answer to the relative roles of government and private R&D. The relative scarcity of private funds—recently put at $5 million annually in the last decade—in itself may be an indication of the uncertainties as to whether shale oil will be commercial within a time horizon that is sufficiently close for corporate planning and decision-making. A careful analysis of alternatives in energy supply will make it possible to assess the alternatives open to government in the field of shale oil research and development, and their implications for the emergence of a privately run, competitive industry.*

Associated Minerals

A recent and wholly new facet of oil shale resources has been the discovery that much of the shale also contains minerals that are potential sources of soda ash and alumina, respectively. Though oil shale was withdrawn from leasing in 1920, there was nothing to prevent claims from being filed for having located either of these minerals. In 1966 alone, some 7,000 claims are estimated to have been filed for dawsonite, the name of the alumina-bearing mineral. These minerals are so intermingled with the oil shale that neither can, it seems, be removed without the other. Thus, oil shale is intimately involved.

There are two aspects to this newly-emerged problem. One is legal, legislative, and administrative, in that a way will have to be found to deal with the validity of these claims through determination of the locatability of the minerals in question or otherwise. This is a matter on which work is now going forward in the Department of the Interior. But the second aspect is an economic one. *Above all, what are the costs of these minerals likely to be? What modifications are called for in the appraisal of shale oil? Should the existence of these minerals affect the ownership and control structure of the shale oil industry, if and when it emerges? What effect might co-production of these minerals have on the cost of shale oil? What would be the effects on current sources of alumina and soda ash? Depending on the stage at which the minerals are separated, how might different approaches to allocating the joint cost affect both shale oil and each of the minerals?*

Undoubtedly, additional questions can be formulated, but it is suggested that a beginning be made with estimating the likely cost and, therefore, competitive standing of the two minerals. It is quite possible that subsidiary issues will then appear less urgently in need of resolution. But as one looks over the literature on the subject, one is struck by the absence of any economic evaluation of these newly prominent minerals. An evaluation of their future markets, in the light of their likely costs, is an important factor in assessing their significance as a complication in shale land valuation and development policy.

VIII

Towards a Co-ordinated Approach to Energy Problems

STUDIES NEEDED TO PROVIDE an informational and analytical base for energy policy relating to the separate industries have been identified in the earlier chapters. But, in view of the extensive and ever-growing interchangeability among the various energy sources and forms, studies confined to the separate industries can yield only a portion of the background knowledge relevant to the formation and evaluation of energy policy. Any effort to direct policy toward co-ordinated objectives will require an understanding of the complicated interrelationships among the separate sources of energy.

We shall present, under two general headings, a number of ways of examining these interrelationships: (1) the quantitative framework for assessing the outlook for energy supply and demand; and (2) energy policies as evaluated from several vantage points, each one representing an area of social concern. In view of its character, this chapter is *not* to be read as a summary of proposals in preceding chapters, but rather as a statement of types of study reaching across the board of the energy industries.

The Quantitative Framework for Assessing Supply and Demand

STATISTICAL PROJECTIONS

A good deal of effort in the postwar years has gone into projecting future energy demands and the adequacy of supplies to meet them. The usefulness of such estimates can be exaggerated, especially if the projection period is overlong. Nevertheless, the disciplined use of projections

129

of future consumption "requirements" for energy—in the aggregate, by types of source, and by types of use—is an essential feature in a program of energy studies. Such projections provide signposts of a sort; by assessing the prospects for future consumption, based on trend analysis, they bring into sharp relief the draft upon fuel supplies that will be presented by rising energy demands.

The fact that projection enterprises flourish in the Executive Branch testifies to the need felt for guidance from them. Thus, the National Power Survey engaged in extensive projections of the demand for electrical energy. Likewise, the proposed National Gas Survey of the Federal Power Commission will presumably prepare projections of the demand for gas. The Atomic Energy Commission maintains a forecasting capability and revises its ideas of future electrical energy demand and its nuclear component with relative frequency. Its forecasts serve to throw light on anticipated developments in uranium demand and problems of supply and on the pace at which various parts of its programs, such as breeder research, might best proceed. Agencies of the Department of the Interior have at various times engaged in projections of energy as a whole as well as portions thereof. One would hardly want to stop the various federal offices from engaging in such work. At the same time, such dispersed activities do not yield an internally consistent picture within the federal establishment. Since projections have substantial uses, the work is best carried on in a co-ordinated manner within the government in order to diminish the inconsistencies inherent in scattered efforts.

To avoid becoming a source of intellectual confusion, projections need to be drawn to clearly defined specifications and accompanied by a precise statement of their meaning and limitations. Today they emanate from many sources and are of different quality, drawn to different specifications, with different coverage; and not infrequently they are designed to promote some particular interest. *Consequently, a function of critical review of projections should exist in an agency of government concerned with general energy problems and policies. But beyond this, and much more important, we recommend that a permanent projections effort be maintained, preferably in the federal government, provided that a carefully constituted research environment can be created.* Parallel efforts by separate federal agencies should not be discouraged. These agencies could contribute to the central effort, and their specialized exercises in projections could be related to it by a code which permitted comparability.

In any plan of consumption projections for energy as a whole, the most difficult problem is that of assigning shares in the total to the different sources. The chief reasons for the difficulty are the highly dynamic state of the energy industries and the existence of many unknown future supply and demand conditions in each of them. It is necessary to make estimates for specific energy sources, in order to permit comparisons with their likely future availability. On the other hand, with the increasing substitutability of one primary source for another, these distinctions are becoming somewhat less important. If, for example, commercially feasible liquefaction and gasification of coal were to be achieved, the use of oil and gas vs. converted coal would become a matter of price differentials in specific locations, and the distinction between the two types of fossil fuels would begin to lose meaning. Similarly—and this trend is already upon us—the expansion of electricity into new uses gives increased prominence to a form of energy that can be derived from a variety of sources. Factors such as these point to the necessity of taking much more account than has been customary in conventional projections of future changes in the technology of production, transportation, and use, and their effects on comparative costs and prices among the different sources.

Energy demand and supply projections would prove more useful if they were drawn up in terms of alternatives, i.e., on varying assumptions with regard to future reserves, technology, and price movements within chosen parameters of economic growth, such as population and GNP. This would not, and probably should not, exclude the preparation of a "most probable" pattern, the variations from which would identify alternative assumptions.

Statistical projections should be built upon careful analysis of the demand-and-supply conditions for each of the various fuels. On the supply side, such analysis is primarily cost analysis, not merely in general terms, but in relation also to geographical markets and to various user markets. In addition to technical cost studies as such, attention has to be given to market structures which condition the pricing practices for the various fuels. (Because of their special importance, industry cost studies are given separate attention in a later section.) On the demand side, analysis must lead to an identification of those uses in which transfers between fuels are most likely to occur in response to changing relative costs. The best foundation for knowledge of this sort would be a detailed study of the fuels consumed in various uses, their interchangeability, and their price-demand elasticity. The data to support such studies would not be complete, but over time they should become sufficient to support very informative results. Background information of the type required points to the need for numerous econometric studies on both the supply and demand side, and studies of industrial organization, particularly in regard to factors affecting costs and prices in the several energy industries.

More than in the past, such studies should take account of the effect that the burgeoning demand of the chemical industry for raw materials might exert on specific portions of energy supply. While in the aggregate this demand now claims only a minor share of energy commodities in terms of volume, trends in petrochemical technology suggest a greatly expanded field of use for substitutes for metallic products. Moreover, petrochemicals may play a greater role in terms of company and industry profits than is suggested by their physical share. The effect of this source of demand on the supply and demand of the parent materials needs greater attention than it has received.

THE PHYSICAL UNDERPINNINGS: "RESOURCES" AND "RESERVES"

A key element in assessing the future adequacy and costs of energy supplies is the natural resource base of the various source materials. This subject may be investigated at three levels:

1. At the level of the widest geological inference concerning the presence of mineral fuel deposits, along the lines of various studies heretofore conducted by the U.S. Geological Survey.

2. At the intermediate level of amounts of source materials which may reasonably be inferred to be forthcoming on economic terms in some fairly extended period of time, such as the next twenty or thirty years, under various assumptions with respect to future costs and prices. This goes beyond estimated recoverable content of known deposits into a limited projection of new discoveries and improved technology of recovery.

3. At the level of recoverable "reserves" estimated to be present in known deposits. At this level, the estimates can be subdivided and classified according to degrees of certainty, and adjusted to different economic and technological assumptions.

In current practice there is a confusing variety of analytical procedures, both with respect to single source materials and as between different materials. *At earlier points in this report, in connection with the separate mineral fuels, the need for studies of reserves and resources has been proposed. At the present point, it is proposed that all such studies should be brought into juxtaposition with one another and that a comprehensive critical review be made of the various analytical procedures used in making reserve-resource estimates.* The importance of this critical cross-comparison arises from the substitutability of the various materials. This fact makes it highly desirable that the estimating procedures give comparable results, insofar as the differing physical characteristics of the deposits of various materials permit.

Drawing upon earlier proposals with respect to particular materials, we may note some elements of the problem of achieving comparability.

Crude Oil and Natural Gas. In the case of crude oil and natural gas, the category of "proved reserves" provides a minimal figure of assured supply from known reservoirs, running to some twelve times the annual rate of consumption in the case of oil and about eighteen for natural gas. American Petroleum Institute (API) and American Gas Association (AGA) figures are the only widely publicized estimates bearing upon future availability; and the strictly defined limits of the "proved" concept tend to overemphasize the scarcity aspect of oil and gas availability. From this base, there are ways of arriving at reasonably expanded estimates of the economically recoverable content of known fields, usefully studied in the past by the Interstate Oil Compact Commission, under various cost and technological assumptions. Estimates of this sort are more important for oil than for natural gas, which has a high recovery rate from known deposits to begin with. From this expanded base, there are further possibilities of statistical predictions of short-run discoveries of reserves in response to degrees of exploratory effort. Beyond this, there is a role for imaginative speculation concerning the longer-run potentialities for oil and gas discovery. Statistical procedures, carried out through time, can connect these various stages of relative certainty and uncertainty. This range of analytical problems was brought into focus by the *Petroleum Statistics Report,* issued in 1965.[1]

Coal. The usual approach to coal starts from precisely the opposite end, from a base which overemphasizes the great abundance of coal. This approach is the consequence of the much greater geological knowledge of the actual incidence of coal in the physical environment, not so much because of greater effort but because, as one of our reviewers put it, ". . . the greater knowledge stems from the nature of the occurrence of the coal—accordingly a little knowledge goes a long way." From the enormous inferred geological base, the Department of the Interior has at times made estimates of amounts minable at close to present costs, running to several hundred times the present annual rate of consumption. This suggests that, in the event of a shortage of other fuels, coal is always there to fill any gaps for a very long time to come.

This mode of estimation does not, however, permit a useful judgment of coal availability in relation to the requirements of the market—such as, for example, putting together large, conveniently located blocks on long-term contract for large electricity-generating stations, or providing

[1] Executive Office of the President, Bureau of the Budget, *Petroleum Statistics Report,* prepared by the Petroleum Statistics Study Group, March 22, 1965 (mimeo).

134   *U.S. ENERGY POLICIES*

sites for minemouth power generation, or providing the materials base for large-scale synthetic oil and gas projects. To bring estimates of coal reserves into the realm of short-run practical reality, for purposes of comparison with other sources of energy, it appears that, as a minimum, new data need to be gathered, or perhaps new methods of estimation need to be devised. One new approach, not necessarily the best but an imaginative step forward, was tried by the National Fuels and Energy Study Group in its 1962 report [2] when, by poll of independent producers, it arrived at an estimate of reserves held by them available at 1960 prices in the amount of 20 billion tons, or forty-seven times the then current rate of consumption, with larger amounts at higher prices. In brief, a methodology is called for by which something corresponding to proved reserves, as in the case of oil, can be connected with the wider bases of geological estimation.

Shale Oil. As the shale oil industry materializes, attention will have to be given to methods of estimating availability within limiting economic and technological assumptions. Here, even more than in the case of coal —and for the same reason, i.e., the character of the resource—geological knowledge is ample; and the enormous oil content of deposits tends to foster erroneous ideas of the relation of shale oil to other sources of energy. Already, indeed before there is an industry, such ideas muddy the discussion of the lines upon which it should be developed. The knowledge of great "ultimate" reserves is a pleasant and very important piece of information to have; but knowledge of availability within defined economic and technical constraints is required for most purposes of policy.

Uranium and Thorium. Given a developed technology for the economic conversion of nuclear fuels to electric energy, the great expectations that have been entertained for nuclear fuels as an abundant source of energy have to be judged in relation to two interrelated factors: (1) the technological efficiency for recovering the heat content of the fissionable materials, and (2) the availability of the raw materials on economic terms. At the present time, uranium is the primary material involved. The AEC has been responsible for developing the estimates of uranium and thorium reserves and resources and has made a commendable start in quantifying resources according to a schedule of estimated costs of bringing them into production. At the same time, there is a great deal of concern about resource availability because of a concentration of

---

[2] U.S. Senate, Committee on Interior and Insular Affairs, *Report of the National Fuels and Energy Study Group on an Assessment of Available Information on Energy in the United States,* September 21, 1962, Senate Document No. 159, 87th Congress, 2nd session (Washington: Government Printing Office, 1962), p. 82.

public attention on statistics of known and estimated well-delineated domestic resources at or near current costs. Something is needed comparable to what is needed in the case of crude oil and natural gas: a method which integrates the information on amounts of known reserves and resources with the information regarding expected future discoveries, estimated according to a cost scale and according to degrees of certainty and uncertainty.

. . .

Across the board, whatever the material under consideration, methods of estimating reserves and resources must contain explicit economic and technological assumptions, as elements in the degree-of-certainty test. "Reserves" of a material must never be thought of as a single "amount," but as a group of amounts, corresponding to alternative assumptions. In the case of crude oil, for example, the volume of "proved" reserves can be thought of as the smallest of a nest of concentric circles, which grow larger as the analytical constraints of technical definition, economic and technological circumstances, and degree of certainty are progressively relaxed. If something similar were devised as an approach to reserves estimation for all the energy materials, a much clearer and more comprehensive picture could be had of the existing and changing reserves situation in relation to the projected requirements for energy in general. *The comparative study of existing methods of reserves estimation, followed by methodological research to place them on a more nearly uniform, or at least comparable, basis should be an important early study effort.*

INDUSTRY COSTS AND SUPPLY FUNCTIONS

In principle, the most important knowledge relevant to the future availability of energy, if it could be obtained, would be a knowledge of the supply functions of the various fuels—that is to say, a schedule of the amounts that would be brought to market at various prices. The nature of the analysis of supply functions is the tracing of the expected course of marginal and average costs, thus defining the prices which would elicit varying volumes of production. Some of the problems and difficulties of analysis of this sort have already been discussed earlier in connection with particular fuels, especially crude oil and shale oil.

Comparative cost analysis, whatever the difficulties and however unsatisfactory the data to support it, is an essential underpinning to any effort to assess the future outlook for energy supply and demand.

Cost problems cast their shadow over every sphere, whether in analysis or public action. Fixing the field prices of natural gas is based upon cost analysis. Restrictions on crude oil imports are related to the costs of domestic production. The costs of uranium and coal will partly determine their use in electrical generation. The costs of oil from shale or of synthetic fuels from coal will condition the terms of their competition with conventional fuels. And so on into every corner of the energy problems with which public policy is concerned.

The present state of affairs is that the cost analysis now available is nowhere near sufficient to meet the needs of either analysis or policy formation. The reasons for this are twofold. The first is that no systematic effort has ever been made to design and pursue a systematic program of cost studies tailored to the needs of economic analysis of the energy industries. Such cost studies as there are have generally been made by industry groups in pursuit of some interest of their own, or by public agencies for the performance of some particular regulatory function. Objective studies, scientifically designed by economists and other industry experts, are rare in number, limited in scope, and based on inadequate data.

The other reason for the deficiency is inherent in the nature of the problem of cost analysis. What is needed is supply functions for the various fuels running some distance into the future. But the future supply conditions of all energy fuels are clouded by a high degree of uncertainty, because what is involved is a probing of the unknown—the costs related to the finding and development of new reserves. This difficulty cannot, however, serve as an excuse for failure to establish a program of cost studies. Some hypotheses concerning the relation between future costs and supplies provide the only economic basis for assessing the outlook for the future of the different energy sources. The proper grounding for these hypotheses is the marshalling of all available evidence. The problems of method break down into stages, moving from short-run to long-run analysis.

We shall not attempt here to describe the various types of relevant cost studies, a long and complicated task. Our basic recommendation is the setting up of a study group to devise a program of such studies. Such a study group would need to include geologists concerned with the estimation of resources and reserves; mining and petroleum engineers concerned with the technology of production and recovery; econometricians concerned with extracting cost functions from the myriad of historical and cross-sectional statistics available for many of the mineral fuels; and economists concerned with the impact of industrial organization on cost and price practices.

Over the whole range of cost analysis, various types of cost studies will be called for. Certain suggestions concerning types of study have been made at earlier points in this report in connection with particular industries. *But the prescriptions for a comprehensive program of such studies, and for critical review of independent studies, will have to be worked out by the technical group assigned to that task. Overall, such studies should be so designed as to permit maximum comparability of cost estimates as between industries, in order to support informed judgments concerning the role of the various fuels in the total energy complex. The ideal would be credible short-run and long-run supply functions for each source of energy. The attainable will fall far short of that; but great improvements in the analytical basis for assessing future outlook and for policy formulation should be possible*

The prices relevant to establishing the competitive balance are the prices at point of final use, only in part derived from prices at point of origin. It is the price of refined fuel oil at the burner point, not the price of crude oil at the wellhead, that affects its power to compete with natural gas for space heating. The prices of gas and coal delivered at the central station establish their competitive status for generating electricity. Back of these final prices lies a sequence of costs—mining, transporting, processing, and distributing—which must be covered by the price. *It is impossible to understand the nature of the cost-price interdependence for each fuel separately without a thorough knowledge of industry structure, joint costs, and pricing practices within each segment of the industries. Such knowledge is necessary as a basis for informed judgment concerning the present and prospective position of the various fuels in competitive markets. For this purpose, students of industrial organization and marketing practice are needed as members of the study group.*

Energy Policies as Evaluated from Several Vantage Points

We turn now to a series of studies intended to evaluate, within a comparative framework, the public policies affecting the various energy industries. These evaluations are designed to be performed from several vantage points, each of them representing an area of social concern or public action closely associated with energy supply and demand. The categories, however, are not self-contained, because social concern gives rise to a variety of public actions designed to serve different goals, but having consequences beyond those specifically intended. It is, therefore, inevitable that parts of the panorama appear under more than one heading. Such overlapping requires mention but not apology. It is in a way the principal reason why a study of this kind is called for in the first place: to sort out the several strands that are interwoven in the fabric of contemporary U.S. energy policy and to subject them to analysis as they affect energy supply and demands as well as one another.

COST-EFFICIENCY OF THE ENERGY INDUSTRIES

The central economic desideratum with respect to energy, in common with all other lines of production, is that whatever the amounts to be supplied they should be supplied at the lowest costs that practical circumstances permit. This principle is applicable in the first instance to the separate supplies of the individual forms of energy. In the end, it applies to achieving the lowest-cost "mix" among the interchangeable forms. This principle of economical operation will not in all circumstances provide a definitive norm for policy. Other considerations, such as national security and concern for the quality of the environment, may provide reasons for deviating from it. But the economic norm needs always to be available; the reasons for deviations from it need always to be subjected to critical examination, in particular the claims of interested groups for preferential treatment.

Our concern is with the manner in which existing public policy affects the efficiency with which the energy industries perform, and with the potentialities for improving efficiency implied by possible changes in public policy. Much of what needs to be studied under this rubric has provided a central focus for several of the research proposals already made in the individual industry chapters, particularly in Oil and Electricity.

The research approach to the study of efficiency which has been outlined in the industry chapters contains a number of common elements. The first stage consists of cost analysis under actual organizational and operational conditions, and with regard to existing public policies where applicable. The second is that of visualizing economies that could be achieved by changes in structures and in policies, given current technology. The third is to introduce the dynamic factor of changing technology in order to visualize the structures of production and of attendant public policies best adapted to economical operation. The central analytical concept is that of devising supply functions under alternative hypothetical conditions.*The operational significance for policy is that of providing evidence concerning lines of public policy that would promote greater efficiency, in the sense of lower costs for separate forms of energy and for energy in general. The type of evidence provided would be the potentialities for lowering costs by changes in public policies and in industrial organization and practice.*

Studies of the type suggested for the separate industries should ideally take their place within the broader framework of studies of the efficiency of the energy industries as a whole. This would be a means of bringing the most basic of all economic aspects to the center of policy attention.

The problems of encouraging industrial cost-efficiency will carry over to the newer energy industries, shale oil and nuclear energy, where government will necessarily play a central role in determining the structure of the industries. Along the competitive margins of all the energy industries, the cost prospects will be a factor in influencing judgment concerning the character and degree of government intervention to stimulate the various potential sources of energy supply, subjects to which we turn in the following sections.

INCENTIVES TO PRIVATE INDUSTRY

The private incentive to engage in productive activity in any field is, of course, prospective profitability. In the ordinary course of American business, the strength of the incentives in particular directions is determined by the market; but they may be affected by government action, either explicitly designed to do so or as a side effect. The incentive structure in the energy industries is affected by public policies in a variety of ways; and the results of policies need to be studied from that point of view, to throw light on the ways in which they provide lures or deterrents to private enterprise.

The widest range of elements affecting incentives is to be found in the policies affecting crude oil; but the exact cumulative consequences are difficult to unravel. Two federal policies, special tax treatment and import restriction, are specifically designed to stimulate investment in the discovery of new reserves, although their effectiveness in this regard needs study, as indicated in Chapter II. Such positive results as they may have in that direction may be neutralized, to a degree difficult to ascertain, by state proration systems. Restrictions on output and inducements to over-investment in reservoir development seem likely to undermine the incentives for exploration.

In one respect, federal policy affects the profitability of natural gas production in the same way as in the case of crude oil, through the percentage depletion tax allowance. In another respect, however, federal policy acts in the opposite direction to limit profitability, through fixing the field price of natural gas. The result of these two factors upon the long-run supply functions for natural gas is highly obscure. The obscurity is increased by the facts (1) that exploration expenditures for oil and for gas are closely associated, (2) that the profits from production in the two fuels largely accrue to the same companies, and (3) that the joint profits may take many other investment directions than exploration for new reserves. *A study attempting to assess the extent to which the policies succeed in promoting the end to which they are purportedly directed, an enhanced future availability of petroleum, would have to delve deeply into the structure of the industry, since the effects of federal policies are so intertwined with state regulatory policies.* (See Chapter II.)

In sharp contrast to oil and gas, public policies relating to bituminous coal have little direct effect upon private incentives, though there is a low federal percentage depletion tax benefit. One reason for this difference, on the side of federal policies, may be that the findings of new reserves is the central concern in the case of petroleum, while great reserves of coal are taken for granted. *The main way in which public policies affect incentives in the coal industry is indirectly, through policies relating to other industries but which affect the position of coal in interfuel competition. This situation could conceivably change if, in addition to its present slight contribution to coal research, the federal government were to mount an extensive program of R&D on synthetic fuels from coal. Leaving aside the competitive repercussions of oil and gas policies, the greatest effects upon private enterprise in coal production are likely to arise out of federal policies with respect to the new sources of energy, nuclear energy and shale oil.*

The roadway to private enterprise in the nuclear field has been provided by the government, through its support of R&D programs, incentives to discover uranium, and subsidies to nuclear power stations. (See Chapter VI.) The future of the coal industry will be deeply affected by the results of these policies and the scope and intensity of their continuation in the future. A critical question of policy will be, as in the past, in connection with government programs in energy R&D.

In the case of shale oil, the pattern of incentives for private operators has yet to be established by government policies with respect to R&D, leasing, taxation, and other decisions which will determine the structure of the industry. If in the end a highly profitable industry for large-scale output emerges, there will be important consequences for the incentive structure in petroleum; and the development of the plentiful coal deposits of the West might be greatly affected.

An urgent field for study, therefore, is the effect of existing and contemplated public policies upon the competitive status of fuels by way of the incentives and disincentives they provide. While, on balance, the competitive position among fuels is to a large degree determined by the operation of private enterprise, the energy industries are much affected by a variety of public policies—production and import control and taxation in the case of crude oil, taxation and price control in the case of natural gas, railroad rate control in the case of coal, public utility rate control in the case of electricity, etc. The competitive status of nuclear energy has been, and will continue to be, strongly affected by government policy and action. The status of shale oil will be established under special rules laid down by government. Regulations with respect to air pollution may greatly affect the markets for the various fuels.

A new factor appearing in the corporate structure of the energy indus-

tries is the acquisition by business firms in one industry of interests in other industries. An example is the acquisition by oil companies of coal properties. On the horizon is the place the great oil companies are likely to seek in the development of the shale oil industry. Conceivably, a large share in the various sources of fuels—oil, natural gas, coal, shale oil, uranium—could fall under the control of such corporate conglomerates, greatly changing the business basis of interfuel competition. The business incentives behind corporate strategy might, or might not, conform to objectives of public policy. *The dynamic features of corporate organization in this direction, and its implications for the impact of the incentives and disincentives contained in governmental policies, will certainly require attention.*

ENERGY RESEARCH AND DEVELOPMENT

The study of the energy industries is made peculiarly difficult by the highly dynamic state of the technology of production and use. The degree to which government R&D should supplement private R&D, and the proportions in which government R&D funds are allocated among the different branches of the industry, have thus come to the fore as a major concern of policy. A critical look at past government programs in support of energy R&D furnishes the reason; it is the episodic and disjointed way in which government R&D in this field has developed, apparently bare of any overriding principle such as might arise out of an overall energy outlook.

The government has in fact taken an active hand in some directions, with telling results. Vast sums have been poured into nuclear R&D (see Chapter VI). By contrast, there was a brief spurt of activity with respect to shale oil a number of years ago, and almost nothing since, and a relatively ambitious program planned for the next ten years. A minuscule program is sponsored with respect to synthetics from coal.

The nuclear program has yielded a method which brings nuclear fuel within the competitive range of coal for purposes of electrical generation. The question presents itself whether, if this were to be the total result, some of the expenditures might better have been spent on alternative programs to develop the technology of shale oil, to stimulate research on the quest for new reserves of oil and gas, or to develop the technologies for turning coal into oil or gas. Barring the successful development of breeder reactors, this is a highly pertinent question. But if the importance of nuclear energy as a great addition to energy supplies depends upon an efficient breeder reactor, the further question arises whether research and development expenditures should not have been concentrated in that direction, and the light water reactors bypassed.

Research efforts in support of an R&D policy less attuned to the opportunities of individual energy sources and more so to energy needs as a whole are appropriate in three directions:

First, there should be a running "inventory" of R&D work in progress and its results. At present, this exists in bits and pieces in the sponsoring agencies, and is sometimes hard to obtain even from them (e.g., work on fuel cells performed by contractors for the several branches of the Armed Forces, NASA, and perhaps others, yet of great potential significance for fossil fuel demand). *A promising start on this has been made by the publication in 1965 of* Energy R&D and National Progress, *prepared by the Energy Study Group in the Executive Office of the President.*[3] *But, as with most efforts made by one-time bodies, the end of the group signifies also the end of the effort.*

Second, there is need for a sorting out of R&D programs, actual and potential, according to carefully selected criteria. These should be partly social (the degree of diffusion of benefits), partly political (national security), partly economic (the prospects for entering the competitive cost range), partly physical (potential quantitative contribution to energy supply), partly financial (estimated costs of an effective program), partly institutional (role of private industry in conducting R&D), and partly temporal (estimated time within which useful results might reasonably be hoped for).

Some such checklist of criteria could serve as the basis for judgment concerning directions and amount of government support of R&D programs. Among those favoring government as compared to private R&D might be:

- A high degree of national security involvement;
- Instances in which social costs and benefits differ widely from private costs and benefits;
- Very large, protracted, or risky enterprises;
- Undertakings that would strengthen competition in the energy industry;
- Research the practical outcome of which is largely affected by regulatory or other policies in the hands of government and from which private industry thus tends to abstain.

A detailed analysis of how the tests cited, and others, might be so used is urgently needed.

[3] Executive Office of the President, Office of Science and Technology, *Energy R&D and National Progress,* prepared for the Interdepartmental Energy Study by the Energy Study Group, under the direction of Ali Bulent Cambel (Washington: Government Printing Office, 1965).

On the basis of agreed-upon criteria, the third line of endeavor would consist of a critical review of present R&D programs.

The outcome of such studies should be the formulation of a rationale for government policy in the field of R&D. The analytical approach to such a rationale is that of cost-benefit analysis. Obviously, in fields where the work to be done and the results to be obtained contain so many uncertainties, the ability to quantify the benefits from outlays in particular directions will be very limited, except on stated assumptions, but at a minimum projects can be judged in terms of short-run urgency and possible long-run contributions to energy supplies. Judgments can also be made concerning the types of R&D that can safely be left to private initiative and those where a large element of government initiative is called for.

The line of thought presented above does not lead to suggestions concerning any particular studies to be conducted in the R&D field, but suggests considerations that would enter such studies. Closely associated is the problem of the type of organizational structure within the government that is to perform the review functions and to provide the advisory judgments.

NATIONAL SECURITY

Because it affects virtually all aspects in the life of the individual as well as the nation, an assured supply of energy, in time of war as well as in time of peace, is an objective of national policy. It has come to be concerned most intimately, however, with the question of oil supplies: To what degree should the country allow itself to become dependent upon foreign sources and how does such dependence affect the prospects of the domestic industry? The oil import program thus became the focus of the debate over the proper role of national security considerations. Security was indeed the sole official reason advanced in justification of it.

The national security objective as it affects oil is discussed in Chapter II. Because of its general importance, however, a few key ideas not dealing directly with any one energy source are repeated here.

First, an inquiry into the national security rationale calls for clarification of the concept itself. A whole congeries of security concepts now invade the sphere of public discussion—a war contingency concept (for different kinds of wars), a national self-sufficiency concept, a Western Hemisphere self-sufficiency concept, a Western Alliance self-sufficency concept, a good foreign relations concept, and finally, though without exhausting the list, a wholeheartedly protectionist concept that what is good for the domestic oil producers is good for national security. These concepts swirl around in the controversial discussion of import policy, creating a general obfuscation of what is being talked about. As a subject for special study, however, national security considerations can be limited to kinds of precautions to be taken to assure energy supplies in a variety of contingencies that might arise in the international situation.

Second, we must ask, what measures are appropriate for meeting what situations? What risks are to be met by what means? This is a subject at present obscured in a fog of undefined ends and means. Several federal, and even state, policies have rested part of their justifications on security considerations. Oil import controls, special tax treatment for oil and gas, federal leasing policy on off-shore oil and gas provinces, programs for coal research, programs for nuclear energy research, and even state conservation regulation have all invoked the national security argument. But no systematic assessment of the security role of energy policies has ever been attempted.

Third, to the extent that national security is the basis for some policies, it is important to know the cost to the economy of this security. As economists have pointed out, to achieve absolute security might require devoting a large share of the nation's resources to security purposes, at the cost of a much lower rate of economic growth and standard of living. It is necessary to determine the trade-off points between higher cost and higher risks. But today scarcely even a framework exists within which to analyze this problem. Moreover, since there are alternative ways of achieving a security goal, the costs of the alternatives need to be analyzed if the national security cost is to be minimized. Cost-benefit analysis has been used in defense systems appraisal. Such an approach is not free from error, but it might well be adapted to studying the security ramifications of alternative energy policies.

FEDERAL PUBLIC LANDS AS AN ENERGY SOURCE

The federal government is placed in a position of peculiar responsibility for future energy supplies by the fact that some of the most prolific potential sources are located in areas under federal ownership or control. This is especially true with respect to oil and gas in the Outer Continental Shelf and shale oil in the Western public lands. To a substantial degree it is true of oil and gas in the Western public lands and of the raw materials for nuclear processes. These various aspects of federal involvement call for a number of different studies.

Potentially the most important subject is that of shale oil. Elsewhere (see Chapter VII) we have outlined studies appropriate to the determination of policy in this field, and need say nothing further here.

With respect to the petroleum resources of the public lands other than the Outer Continental Shelf, there appear to be good reasons (as indicated in Chapter II) for a critical review of the whole framework of present leasing policy. Questions have been raised concerning it that deserve answers: about how far it serves as a vehicle for speculative trading in leases; how far it encourages the postponement of exploratory effort; and how far it fails to require the most efficient methods of reservoir development.

A more specific and potentially more far-reaching question can be raised concerning federal leasing and administrative practices on public lands including the Outer Continental Shelf, in relation to the regulatory practices of the states. This subject is dealt with in Chapter II, but is referred to again here because of its critical importance. A major problem of the oil-producing industry has been the inefficiencies resulting from state regulatory practices; in the future, rising costs and declining incentives to explore for new reserves could make improvement in this situation imperative. *The federal government is in a position where it could, if it so decided, introduce a code for oil production which would serve as a model for state action. The federal lands could provide a laboratory to see how an effective unitization program could be set up for new fields as well as old ones. In doing so, production on federal lands would have to be dissociated from the operation of state proration systems—to which it now conforms, not by legal necessity but at the option of the government. Present policy is to support the operation of the state systems. The alternative would be to demonstrate the methods by which they could be usefully modified. A preliminary study could bring out the problems to be encountered in such a change of policy.*

In another direction, a study could be made of the ways in which oil on public lands, including the Outer Continental Shelf, could be fitted into plans for national security. A problem of the future may be foreshadowed by the declining reserves-production ratio, though a better understanding of the meaning of this ratio would be needed before it could be used as an indication of stringency in supply. The government could, by its own enterprise or by incentives to private enterprise, stimulate the discovery and partial development of fields, but hold them out of production. Public lands, including offshore, now account for 10 per cent or more of national output; and this is growing. Some part of the present reserves and a larger part of new reserves could be kept under wraps and set aside as a national stockpile. (As a precedent, the past use—or nonuse—of the Naval Petroleum Reserve in Elk Hills, California, deserves study in this context, as does the Navy's promotion of oil shale development.) The consideration of any such plan would have to be in the broader context of alternative measures for assuring energy supplies in the interest of national security.

The broad range of policy questions affecting mineral fuels on public lands is now being reviewed by the Public Land Law Review Commission. The Commission's efforts in this regard will go well beyond the specific questions identified here. The subjects that we have suggested for study within an energy research program are included solely because of their strong interrelationship with other significant policy matters in the field of energy that have been identified in this report.

IMPACT ON THE NATURAL ENVIRONMENT

In recent years the adverse effect on the natural environment of effluent-polluted air and water and of despoiled landscape has become a matter of urgent public concern. Although covering a much wider range of phenomena, its relevance for the different branches of the energy industry is increasing in importance. The adverse effects are of two orders: those affecting health and those affecting the usability and beauty of the physical environment.

In the absence of pollution controls, the direct costs assumed by business leave out of account the social costs of environmental damage. The problems of policy are those of deciding the extent to which the benefits derived from reducing deleterious consequences justify the costs of doing so. The size of the costs imposed upon private business and the ways in which these costs are distributed among firms and industries will affect the operations and competitive situations of the several energy industries.

Pollution of streams caused by oil production and to some extent by acid effluents from coal mines, as well as the damage done to soil by strip mines, have long been problems for the affected states and subject to legislation. So have some other noxious phenomena. But in general, and especially in combating air pollution, governmental policy at federal, state, and local levels is still in a formative state. Of particular significance for energy matters was the Clean Air Act of 1963, since amended in various ways, which was the first national attempt to assist in advancing the state of knowledge, the state of the arts in dealing with the phenomenon, and the state of institutions administering the law.

Insistent public pressure has brought about some quick responses the wisdom of which should be tested in the light of information yet to be acquired. The first need, therefore, is for research to specify the kinds of data required to support judgments on policy. Together with state and local laws and regulations, federal legislation affects energy use in two different contexts: (1) as a means of highway propulsion, and (2) as a source of stationary heat and power.

In highway propulsion, under the 1965 amendments to the Act, standards of emission are laid down for various pollutants, the achievement of which requires modification of the automobile engine and associated equipment. It may in the future also require modification of the fuel. So long as no acceptable substitute for the gasoline or diesel motor has been developed, measures to reduce noxious emissions would pose no threat to the competitive position of oil. They would simply impose some extra costs on users. However, the air pollution problem has created a renewed interest in the possibilities of electrically powered automobiles. Any such development on a significant scale appears to be distant, depending above all on innovations in the construction of batteries; but it cannot be counted out as a factor in long-run energy supply. While studies have been made of the conditions of emergence and the spread of an electric automobile—or some other alternative, such as a steam-driven vehicle— less attention has been devoted to the likely impact upon energy demand and supply. Such study is indicated, even in the absence of an immediate appearance of the electric vehicle.

In the case of air pollution from *stationary sources*, such as coal and oil for electric power plants, space heating, and industrial use, the problems of imposing standards are more complex because of the diverse situations in which the pollution occurs. This, moreover, is the context within which public policies may substantially affect the choice of sources of supply and the competitive position of the various fuels. They have already begun to do so in a few instances.

The core of present policy is one of laying down emission standards and leaving energy users and producers to make the necessary adjustments and bear the resulting costs. To determine and evaluate both "necessary adjustments" and "resulting cost," however, runs into great difficulties. Areas of major ignorance comprise principally (1) the long-run effects, on both animate and inanimate matter, of exposure to sustained low levels of energy-associated pollutants; (2) the expression of such effects in monetary units; (3) the costs (both private and social, direct and indirect) of decreasing the pollutant content of fuels or substituting one fuel for another; (4) the trade-offs between controlling pollution at the source through selection and/or treatment of the fuel, or at the point of emission; and (5) relating such costs to the benefits of suppressing or ameliorating the adverse effects under (1) shown above.

The phases of policy most likely to pose early and serious problems for the energy industries are those concerned with air pollutants—especially from oxides of sulfur—caused by combustion of oil and coal in power plants and in other industrial and space heating uses. Legislation takes the form of specifying emission control levels and regional air quality standards. The additional cost of de-sulfurizing could become a deciding factor in interfuel competition in areas in which the competitive margins are thin. Efforts to control pollution give an inside track to natural gas which is a relatively "clean" fuel. They also favor sources of oil and coal that have a low sulfur content. They run against the traditional supplies of imported residual fuel oil to users along much of the East Coast. They could become an important factor in the rate at which nuclear energy displaces coal in the generation of electricity.

Short-run dislocation could be severe, especially if stringent standards are imposed by law prior to the development of efficient technologies of purification. Against this must be weighed the consideration that a certain degree of stringency is itself an element in advancing the momentum of research and development.

This is about where the matter now stands. Coal producers and electric power companies are worried about long-run contracts for coal supplies conforming to as yet undefined emission and air quality standards. Suppliers and users of imported residual fuel oils are subject to similar uncertainties, as are suppliers and users of domestic fuel oils for space heating.

The search for satisfactory sources of fuels is spreading; research into anti-pollution expedients is going on at many points, as is consideration of appropriate public policies at federal, state, and local levels. *Those charged with general responsibility for keeping the energy situation under review will necessarily have to keep in mind certain questions for which answers should be sought, as illustrated below in relation to coal, oil, and natural gas.*

Coal

1. *How large and where located is that part of the market for coal that would be affected by limitations on sulfur content?*

2. *What is the magnitude of the reserves of low-sulfur coal in the United States? What other characteristics are associated with low sulfur that might affect economics of combustion? What are the circumstances of ownership, and to what extent are these deposits dedicated to specific uses? What are the costs at which increasing quantities of such premium sources would become available?*

3. *What are the costs, per million Btu, of alternative methods of reducing coal's sulfur content prior to combustion? What by-products are yielded by each of the alternatives and what are the most realistic assumptions that can be made as to the financial return on their disposal?*

4. *What are the costs, per million Btu, of alternative methods of reducing emission of sulfur oxides after combustion? How do such costs vary with degree of removal? What is the nature and possible financial return from disposal of material captured in the course of emission?*

5. *Can the quality of ambient air be used as the standard in place of sulfur content of fuel or stack gas? If so, how would costs and administrative feasibility compare with (3) and (4) above?*

Oil

The questions arising in the case of oil would in general be similar to those listed for coal, adapted to the special markets in which the fuel oils are utilized, and the sources from which the material arises, including the various foreign sources of supply.

Natural Gas

1. *What are the long-term implications for natural gas supply of substituting natural gas for coal and oil? Is there reason for concern over the consequent increased draft on gas reserves?*

2. *Under the Natural Gas Act of 1938, responsibility for regulating the transportation and sale for resale in interstate commerce of natural gas is vested in the Federal Power Commission. This regulatory power is crucial. It can grant or withhold permission to ship natural gas into specific areas or commit it to specific uses. With increasing frequency, the Commission has been faced with the question of what criteria it should employ in its decisions regarding gas deliveries demanded wholly or in part on grounds of air pollution control. Do the non-polluting characteristics of gas call for a reversal of the conventional view of boiler fuel as an "inferior" use? What would be the regulatory consequences of such a reversal?*

   The issues that arise relate to both substance and procedure, and include analysis (a) of the effects of increased gas deliveries on gas costs (in all markets), gas reserves, and, of course, on the extent of pollution abatement; and (b) of the institutional channels through which decisions by the Federal Power Commission concerning gas deliveries are reached, specifically, of the extent to which agencies outside the FPC can and should be party to the decisions.

ASPECTS OF SOCIAL WELFARE

One point of view from which interfuel competition and the dynamics of change should be studied is that of the welfare of the people and communities dependent on the industries. The people most directly in the line of fire are those in the coal mining industry. Throughout most of its history, the industry has been beset by problems of excess labor supply, chaotic competitive conditions, inefficient productive organization, and disturbed labor conditions. These conditions were exacerbated by the Great Depression and the competitive impact of petroleum. In the post-war years the industry has protected its competitive position by a cost-reducing program of mechanization with the collateral result of a drastic reduction of manpower requirements. *This has been an important element in the problems of Appalachia and has inspired programs for bringing new sources of livelihood to the region and for shifting the excess labor force.*

This situation might have been capable of a long-run solution; though not without prolonged agony to many communities and great outlays on welfare and development programs. However, there are now new threats to the coal industry because of nuclear energy, and the prospects for a severe anti-pollution program. Offsetting this threat is the rapidly mounting demand for energy in general, accompanied by advances in the technology of coal utilization. These possibilities generate different degrees of optimism and pessimism concerning the prospects for mining communities and regions. On the pessimistic side, within the industry one encounters a widespread view that the scales are heavily weighted against coal by the failure of the government to mount an R&D program on coal conversion comparable to the nuclear energy program. With the added threat of anti-pollution measures prior to development of appropriate technologies of control, many foresee a crisis of major dimensions for coal within the next twenty years.

The welfare implications of these problems certainly ought to be brought within the field of vision. Studies should be initiated which will utilize the skills of social scientists who will need to become familiar with the problems of the coal country; with the future prospects implied in technological research in energy R&D programs, and in anti-pollution measures; and with studies of the changing industrial and market organization of the coal industry

In principle, severe anti-pollution measures might be expected to have adverse effects upon some oil-producing communities, but in no way comparable to the case of coal. In view of the increasing requirements for oil, it seems unlikely that such measures would have much effect on the rate of production even of the more sulfur-laden varieties. In the case of fuel oils, they would impose some costs of purification and favor natural gas to the extent that it was available as a substitute. In the case of motor fuels, they would add something to the fuel cost of transport. It is difficult, however, to think that these effects would have any very extensive impact upon the welfare of oil-producing communities, though they might have some mildly limiting consequences for the growth of the industry.

There are ways in which a change of policy could raise welfare problems of importance for oil-producing communities. An example would be the elimination or radical relaxation of oil import restrictions. The price effects of this would make much oil production unprofitable. Traced back to the grassroots, this would mean that numberless communities and their members would be greatly damaged. In contrast to the coal country, where the fortunes of the participants have been left to the buffeting of competitive forces, the welfare of the oil communities has been a central consideration in state regulatory action. Nevertheless, in the broader context of policies concerned with national security and economic growth, it is possible that in time the issue of relaxation of import controls might arise. If so, the welfare issue might also arise.

. . .

This report has been limited to proposing a program of informational and analytical studies that would lead to an improved understanding of the energy policies now at work in the United States or under serious discussion. The policies applicable to the separate energy industries arose out of circumstances specific to each of them. An attempt has been made to describe some of these circumstances.

Among these policies there is no common element allying them all to an "energy policy" designed to fulfill clearly defined social objectives. In part, our proposals refer to the problems of individual industries. More importantly, however, and particularly in this final chapter, they cut across the whole spectrum of industries, a procedure necessitated by the increasing interdependence of the sources of energy.

The nature of an "energy policy" is in some sense to co-ordinate the policies relating to separate sources and industries, by modifying them and supplementing them in ways that would serve the broader objectives to which they all should contribute. But "co-ordinate" is a sweet-sounding word that tends to cover up the consequences for many of the interests affected if the government should conscientiously attempt to frame a generic energy policy. It also tends to cover up the complexity of the decisions that government would have to make.

The present report goes no further than proposing a program of special studies which would, in their cumulative effect, provide an analytical overview of the interlaced situations that confront a co-ordinated approach to energy policy. However, the organization of such a program, designed to provide unbiased analysis on detailed topics, will not be a simple matter. Research, some of it reflected in this report, has already been performed on many of the subjects that have been identified. The results of such research will need to be assembled and evaluated. But much more research will be required if the agenda presented in the report is to be adequately covered. Priorities among subjects will need to be established. Finally, given the existence of an ongoing program that begins to yield useful results, the next stage would be organization for performing the difficult task of winnowing the material, or distilling the essence, in a manner to support rational policy decisions in a changing environment.